CHEMOTHERAPY AND AQUATIC THERAPEUTICS

The book entitled "Chemotherapy and Aquatic Therapeutics" deals with different kinds of chemotherapeutants that can be used in the treatment of diseases affecting fish. The mechanism of action behind every therapeutic agent is explained clearly for a better understanding of the basics of the drugs. Effective treatment would be achieved by proper delivery of the compounds at the right time. Different drug delivery methods to be practiced on farm are also deliberated in detail. This book will be immensely helpful to the fisheries students at the undergraduate and post graduate level and scholars pursuing research in the area of aquatic animal health management.

Dr. A. Uma is working as Professor in Tamilnadu Dr. J. Jayalalaithaa Fisheries University and Head of the Department of Fish Pharmacology and Toxicology, Institute of Fisheries Post Graduate Studies. She has specialized in Aquatic animal health management with twenty years of experience in teaching and research. She has guided students for their research and has served as an expert in the doctoral committee of several research students.

Ms. A. Angela Mercy is currently working as Teaching Assistant in the Department of Fisheries Biotechnology at Institute of Fisheries Post Graduate Studies, a constituent unit of Tamil Nadu Dr. J. Jayalalithaa Fisheries University. She is a recipient of Junior Research Fellow from Indian Council of Agricultural Research, New Delhi to pursue her post-graduation at Central Institute of Fisheries Education, Mumbai.

Dr. K. Karal Marx is basically a Fisheries Biotechnologist mainly involved in teaching of courses on "Fish genetics and breeding" and "Fisheries Biotechnology". He has having more than 28 years of teaching and research experience. He has worked as Professor & Head, Department of Fisheries Biotechnology at Fisheries College and Research Institute, Thoothukudi.

CHEMOTHERAPY AND AQUATIC THERAPEUTICS

A. Uma
Professor and Head
Department of Fish Pharmacology and Toxicology
Institute of Fisheries Post graduate studies
Tamil Nadu Dr. J. Jayalalithaa Fisheries University
OMR Campus, CHENNAI-603103

A. Angela Mercy
Teaching Assistant
Institute of Fisheries Post graduate studies
Tamil Nadu Dr. J. Jayalalithaa Fisheries University
OMR Campus, CHENNAI-603103

K. Karal Marx
Dean
Institute of Fisheries Post graduate studies
Tamil Nadu Dr. J. Jayalalithaa Fisheries University
OMR Campus, Chennai-603103

CRC Press
Taylor & Francis Group
Boca Raton London New York

CRC Press is an imprint of the
Taylor & Francis Group, an **informa** business

NARENDRA PUBLISHING HOUSE
DELHI (INDIA)

First published 2021
by CRC Press
2 Park Square, Milton Park, Abingdon, Oxon, OX14 4RN

and by CRC Press
6000 Broken Sound Parkway NW, Suite 300, Boca Raton, FL 33487-2742

© 2021 Narendra Publishing House

CRC Press is an imprint of Informa UK Limited

Print edition not for sale in South Asia (India, Sri Lanka, Nepal, Bangladesh, Pakistan or Bhutan).

British Library Cataloguing-in-Publication Data
A catalogue record for this book is available from the British Library

Library of Congress Cataloging-in-Publication Data
A catalogue entry has been requested

ISBN: 978-0-367-62282-4 (hbk)
ISBN: 978-1-003-10870-2 (ebk)

**NARENDRA PUBLISHING
HOUSE**

Contents

Preface

Chemotherapy is the treatment of disease by the use of chemical substances and drugs. As in animals and humans, chemotherapy finds an important place in treating the diseases infecting fish. This book has been prepared with the basic information on various chemicals and drugs that could be used for treating fish diseases and their dosages for effective treatment of diseases in fish.

The book consists of twenty chapters on various aspects of chemotherapy and aquatic therapeutics. The chapter on general principles of chemotherapy gives an overview of the concepts in chemotherapy. The history of drug discovery and development and the various chemicals and drugs used in aquaculture are given in detail. The methods of sample collection and administration of substances in fish are explained with pictorial representations in an easy and understandable way.

A.Uma
A. Angela Mercy
and
K. Karal Marx

1

INTRODUCTION

Aquaculture is growing rapidly in many regions of the world, and aquaculture products constitute an important food supply with increasing economic importance. Few decades ago, few species were cultured in India mainly in inland waters. In the last 10-15 years, there is a rapid growth and expansion in aquaculture with significant contribution to total fish production. The area under aquaculture is increasing steadily in inland and coastal brackish water. Intensification of culture activity with high stocking density, artificial feed and fertilizers has resulted in the change in water quality parameters and microbial load and has significant negative impact on the health of fish in culture system. Disease is a component of the overall welfare of fish (Bergh, 2007). To control infectious diseases, use of antimicrobial agents is followed in aquaculture as in other areas of animal production. Many diseases can be cured by chemotherapy i.e. use of chemicals that selectively inhibit or kill the pathogens without killing the animal. Drug treatment of diseases caused by invading organisms like bacteria and the other pathogenic microorganisms, parasites, and tumor cells is referred to as chemotherapy. Drugs in chemotherapy differ from all others in that they are designed to inhibit or kill infecting organism and to have minimal effect on the recipient. The objective of chemotherapy is to study and to apply the drugs that have highly selective toxicity to the pathogenic microorganisms in host body and have no or less toxicity to the host, so as to prevent and cure infective diseases caused by pathogens. Thus, chemotherapy involves treatment of systemic infections with specific drugs that selectively suppress the infecting microorganism without significantly affecting the host.

Chemotherapy has been applied in aquaculture for over 60 years since there were many outbreaks of disease in the growing industry. A wide range of chemicals are used in aquaculture, including antibacterials, pesticides, hormones, anaesthetics, various pigments, minerals and vitamins, although not all of them are antibacterial agents. The discovery of antimicrobial agents changed the treatment of infectious diseases, leading to a dramatic reduction in morbidity and mortality, and contributing to significant advances in the health of the general population. In aquaculture, antimicrobial agents are used both prophylactically, at times of heightened risk of disease and therapeutically, when an outbreak of disease occurs in the system. Prophylactic treatments are mostly confined to the hatchery, the juvenile or larval stages of aquatic animal production. There are no antibacterial agents that have been specifically developed for aquacultural use.

Categories of chemotherapeutic Agents

Chemotherapeutic drugs are broadly divided into following classes.

1. Antibacterial drugs
2. Antitubercular drugs
3. Antileprotic drugs
4. Antiviral drugs
5. Antifungal drugs
6. Antiprotozoal drugs
7. Antihelmintic drugs (anthelmintics)
8. Antineoplastic drugs

Initially the term chemotherapeutic agent was restricted to synthetic compounds, but now since many antibiotics and their analogues have been synthesized, this criteria has become irrelevant; both synthetic and microbiologically produced drugs need to be included together. However, it

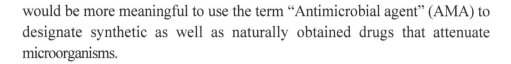

would be more meaningful to use the term "Antimicrobial agent" (AMA) to designate synthetic as well as naturally obtained drugs that attenuate microorganisms.

History of chemotherapy

The history of chemotherapy may be divided into 3 phases.

The period of empirical use (1890-1935): During this period many traditional practices were used for the treatment of diseases. Mouldy curd by Chinese on boils, chaulmoogra oil by hindus in leprosy, chenopodium by Aztecs for intestinal worms, mercury by Paracelsus (16th century) for syphilis, cinchona bark (17th century) for fevers are few examples. The phenomenon of antibiosis was introduced by pasteur in 1877.

Enrlich's phase of dyes and organo metallic compounds (1890-1935): Discovery of microbes occured during this period. In the latter half of 19th century microbes were identified to be the cause of many diseases. Ehrlich proposed the idea that if certain dyes could selectively stain microbes, they could also be selectively toxic to these organisms, and tried methylene blue, trypan red etc. He coined the term chemotherapy because he used drugs of known chemical structure (that of most other drugs in use at that time was not known) and showed that selective attenuation of infecting parasite was a practical proposition.

The modern era chemotherapy: Domagk in 1935 demonstrated the therapeutic effect of a sulphonamide dye, in pyogenic infection. It was soon realized that the active moiety was para amino benzene sulphonamide, and the dye part was not essential. Sulfaphridine was the first sulphonamide to be marketed in 1938.

Fleming (1929) found that a diffusible substance was elaborated by Pencillium mould which could destroy Staphylococcus on the culture plate. He named this substance as penicillin but could not purify it. Chain and Florey followed up this observation in 1939 which culminated in the clinical use of penicillin in 1941. Because of the great potential of this discovery in treating war wounds, commercial manufacture of penicillin soon started.

In the 1940s, Waksman and his colleagues undertook a systematic search of Actinomycetes as source of antibiotics and discovered streptomycin in 1944. This group of soil microbes proved to be a treasure (house of antibiotics) and soon tetracycline, chloramphenicol, erythromycin and many others followed. All three groups of scientists Domagk, Fleming–Chain– Florey and Waksman received Nobel Prize for their discoveries. In the past 40 years, emphasis has shifted from searching new antibiotics producing organisms to developing semi-synthetic derivatives of older antibiotics with more desirable properties or differing spectrum of activity. Few novel synthetic AMAs have also been produced.

Milestones

1929   Penicillin was discovered in England

1932   Sulphonamide (Prontosil) was discovered in Germany

1937   1st sulphonamide released

1938   Serious infections responded to sulphonamides.

1939   Gramicidin was discovered in U.S.

1940   Florey demonstrated penicillin's effectiveness.

1942   Penicillin introduced in England & U.S.

1943   Streptomycin discovered in U.S.

1943   Bacitracin was discovered in U.S.

1945   Cephalosporin was discovered in Italy

1947   Chloramphenicol was discovered in U.S.

1947   Chlortetracycline was discovered in U.S.

1949   Neomycin was discovered in U.S.

1950   Oxytetracycline was discovered in U.S.

1952   Erythromycin was discovered in U.S.

1954   Penicillin-resistant infections become clinically significant

1956   Vancomycin was discovered in U.S.

1957   Kanamycin was discovered in Japan

1960   Methicillin was introduced in England & U.S.

1961   Ampicillin was introduced in England

1963   Gentamicin was discovered in U.S.

1964   Cephalosporins was introduced in England

1966   Doxycycline was introduced in U.S.

1967   Clindamycin was reported in U.S.

1968   Gentamicin-resistant pseudomonas and methicillin-resistant staphylococcal infections became clinically significant.

1969   Amikacin was discovered

1970   In Early 70s, increasing trend of nosocomial infections due to opportunistic pathogens

1971   Tobramycin was discovered in U.S.

1972   Cephamycins was (cefoxitin) discovered in U.S.

1972   Minocycline was introduced in U.S.

1973   Carbenicillin introduced

1974   Ampicillin-resistant infections become frequent

1978   Expanded spectrum cephalosporin Cefoxitin was introduced

1979   Oral cephalosporin with improved activity and Cefaclor were introduced

1981   Anti-pseudomonal cephalosporin Cefotaxime was introduced

1983   Clavulanic acid-amoxycillin was introduced

1985   Norfloxacin was introduced

Fishes suffer from environmental, nutritional, viral, bacterial, parasitic, and neoplastic diseases many of which are similar to those of higher animals. The prevention and treatment of these diseases follow the same principles as diseases of other animals. Many of the drugs and chemicals used in chemotherapy of fishes are the same as for higher animals. There are many antibacterial drugs for animal health. It is well recognized that the issues relating to antibacterial use in animal food are of global concern. Currently, there is a general perception that veterinary medicines, and in particular antibacterials, have not always been used in a responsible manner. In some cases, rather than providing a solution, chemotherapy may complicate health management by triggering toxicity, resistance, residues and occasionally public health and environmental consequences. As a result, authorities have introduced national regulations on the use of antibacterials.

2

GENERAL PRINCIPLES OF CHEMOTHERAPY

C hemotherapy is a branch of pharmacology dealing with drugs that selectively inhibit or destroy specific agents of diseases like bacteria, viruses, fungi, and parasites. The term chemotherapy is also extended to the use of drugs in the treatment of neoplastic diseases due to analogy between the malignant cells and the pathogenic microbes. Paul Ehrlich first coined the term chemotherapy in 1913 to describe drugs that attack invading organisms without harming the host. Thus, the notion of selective toxicity is central to chemotherapy. As a class, the chemotherapeutic agents are one of the most frequently used as well as misused drugs.

TERMS AND CONCEPTS

1. Agents

Chemotherapeutic agents: Chemotherapeutic agents are the drugs used in chemotherapy to interfere with the functioning of foreign cells. These include both antibiotics and synthetic antimicrobials and cover all antibacterial, antifungal, antiviral, anti-protozoal, anthelmintic, and antineoplastic drugs.

Antibiotics: Antibiotics are chemical substances produced by various species of microorganisms (e.g., bacteria, fungi, actinomycetes) that kill or suppress growth of other microorganisms. Common usage often extends the term antibiotic to include synthetic antimicrobial compounds (e.g., sulphonamides and quinolones), but in strictest sense they are not antibiotics, but antimicrobials.

Antimicrobials: Antimicrobials are all chemical substances whether natural, synthetic or semi-synthetic, that kill or suppress the growth of microorganisms.

2. Selective toxicity

Selective toxicity is the ability of an antimicrobial agent to kill an invading microorganism without harming the cells of the host. In most instances, the selective toxicity is relative, rather than absolute, requiring that the concentration of antimicrobial be carefully controlled to attack the invading organisms while still being tolerated by the host.

3. Antimicrobial activity

Antimicrobial activity refers to the ability of a compound to react with the microbial cell molecules in a way that interferes with growth and multiplication of the microorganisms (static effect) or causes killing of the microorganism (cidal effect). Accordingly, depending on the types of organism the activity could be bacteriostatic/bactericidal, fungistatic/fungicidal, amoebistatic/amoebicidal, etc.

Bacteriostatic activity: It is the ability of an antibacterial agent to inhibit the growth and multiplication of bacteria. The inhibited growth in time results in death of the organism and/or removal of organism by the host's defence cells.

Bactericidal activity: It is the ability of an antibacterial agent to cause the death of bacteria. The distinction between static and cidal actions is narrow for certain drugs as some bacteriostatic drugs become bactericidal at higher concentrations (e.g., erythromycin, nitrofurantoin) and some bactericidal drugs are only bacteriostatic under certain circumstances (e.g., streptomycin). It is also possible for an antimicrobial agent to be static for one organism and cidal for another e.g., chloramphenicol is bacteriostatic against Gram-negative rods and bactericidal for Pneumococci.

4. Antimicrobial (Antibacterial) spectrum

Antimicrobial spectrum refers to the range of pathogenic organisms (e.g., bacteria) against which an antimicrobial agent is active. Antimicrobials are often classified as broad-spectrum or narrow-spectrum. The broad-spectrum antimicrobials are active against a wide variety of organisms (e.g., both gram-positive and gram-negative bacteria), while narrow range antimicrobials are active against a few or a class/type of organisms (e.g., gram-negative or gram-positive bacteria). Extended spectrum is the term applied to antimicrobials that are effective against gram-positive bacteria and also against a significant number of gram-negative bacteria (e.g., amplicillin).

5. Potency

Potency may be defined as the antimicrobial activity per milligram (microgram) of a chemotherapeutic agent. Potency is usually expressed on the basis of minimum inhibitory concentration, minimum bactericidal concentration or minimum antibiotic concentration. All these indices are determined *in vitro*.

Minimum inhibitory concentration (MIC): It is the lowest concentration of an antimicrobial drug that prevents visible growth of bacteria when grown against sequentially diminishing drug concentrations *in vitro*. The *in vitro* MIC of the drug guides selection of drugs that can reach similar concentrations *in vivo* and provides a basis for comparing the relative susceptibility of the organism to other drugs. Although the MIC is the most commonly used standard by which the activity of a particular antimicrobial agent is judged, the reported MIC for a particular bacterial species is not always constant. The MIC is affected by host factors, methodology used to determine MIC, different bacterial strains prevalent in a region, etc. The MIC of bacteria may differ with a subsequent infection by the same bacteria and also may change during the course of infection.

MIC90: It is the MIC necessary to inhibit 90 percent of the organisms tested.

Minimum bactericidal concentration (MBC): It is the lowest concentration of an antimicrobial drug that kills the bacteria. The MBC is also determined *in vitro* in a fashion similar to the MIC.

6. Other terms

Post-antibiotic effect (PAE): It is the persistence of the antimicrobial effect for longer period (few hours) after brief exposure to or in absence of detectable concentration of an antimicrobial drug. The post-antibiotic effects vary with each drug and each organism. The PAE can affect the dosing interval for some antimicrobials (e.g., aminoglycosides are given at 12 to 24 hours intervals, although their half-lives are much shorter).

Biphasic ('Eagle') effect: It is phenomenon in which low doses of an antibacterial *in vitro* against certain bacteria (e.g., staphylococci and streptococci) produce lysis of the organism whereas high dose do not. The biphasic effect is associated primarily with beta-lactam antibiotics and is believed to be due to the differential sensitivity of the penicillin binding proteins to high doses of β-lactam that inhibit the autolysins.

ANTIMICROBIAL AGENTS

Properties of an Ideal Antimicrobial Agent

An ideal antimicrobial agent (AMA) should possess following characteristics:

a) It should have a powerful action against microorganisms.

b) It should be specific in action i.e., acts specifically on invading organisms without any toxicity to host.

c) It should not be inactivated rapidly by tissue enzymes or GI microflora.

d) It should have good oral bioavailability and penetrate efficiently to various body tissues and fluids.

e) It should have long elimination half-life and not rapidly excreted by kidneys/bile.

f) It should not favour bacterial resistance and show cross-resistance with other antimicrobial agents.

g) It should not interfere with host immune mechanisms.

h) It should not show adverse drug interactions with other antimicrobial drugs.

i)   It should have no/short withdrawal time in food-producing animals.

j)   It should be easily available and cheap.

Classification of antimicrobial drugs

The antimicrobial drugs can be classified in several ways depending on their mechanism of action, chemical structure, types of organism affected, antimicrobial spectrum, type of action, source, etc.

A. Mechanism of action

1. Agents that inhibit cell wall synthesis

    e.g., Penicillins, Cephalosporins, Cycloserine, Bacitracin, Vancomycin and Clotrimazole.

2. Agents that inhibit cytoplasmic membrane function

    e.g., Polymyxins, Amphotericin B and Nystatin.

3. Agents that inhibit protein synthesis

    e.g., Chloramphenicol, Tetracylines, Macrolides and Aminoglycosides.

4. Agents that affect nucleic acid metabolism and synthesis

    e.g., Quinolones, Rifampicin, Idoxuridine and Acyclovir.

5. Agents that interfere with intermediary metabolism

    e.g., Sulponamides, Trimethoprim and Sulphones.

B. Chemical structure

1. Sulphonamides

    e.g., Sulphadimidine, Sulphadiazine, Sulphanilamide and Sulphaquin-oxaline.

2. Diaminopyrimidines

    e.g., Trimethoprim, Ormetoprim and Baquiloprim.

3. Quinolones

    e.g., Nalidixic Acid, Enrofloxacin, Difloxacin and Ciprofloxacin.

4. Beta-lactam antibiotics

    e.g., Penicillin G, Ampicillin, Cloxacillin, Cefazolin and Cephalexin.

5. Aminoglycosides

   e.g., Streptomycin, Gentamicin, Amikacin and Tobramycin.

6. Tetracyclines

   e.g., Oxytetracycline, Tetracycline, Doxycycline and Minocycline.

7. Macrolide antibiotics

   e.g., Erythromycin and Azithromycin.

8. Polypeptide antibiotics

   e.g., Polymyxin B, Colistin and Bacitracin.

9. Nitrofuran derivatives

   e.g., Nitrofurantoin and Furazolidone.

10. Nitroimidazoles

    e.g., Metronidazole and Tinidazole

11. Polyene antibiotics

    e.g., Nystatin and Amphotericin-B.

12. Imidazole derivatives

    e.g., Ketoconazole, Fluconazole and Clotrimazole.

C. Type of organism/Therapeutic use

1. Antibacterials

   e.g., Penicillins, Aminoglycosides, Tetracylines and Chloramphenicol.

2. Antifungals

   e.g., Amphotericin B, Grisseofulvin and Ketoconazole.

3. Antivirals

   e.g., Idoxuridine, Vidarabine, Zidovudine and Ribavirin.

4. Antiprotozoals

   e.g., Metronidazole, Quinapyramine and Diminazine.

5. Anthelmintics

   e.g., Albendazole, Levamisole, Niclosamide and Praziquantel.

6. Ectoparasitcides

   e.g., Cypermethrin, Lindane, Amitraz and Ethion.

D. Spectrum of Activity

1. Narrow spectrum antimicrobials

   e.g., Penicilin G, Streptomycin, Erythromycin and Vancomycin

2. Broad spectrum antimicrobials

   Tetracycline, Chloramphenicol, Cephalexin, Gentamicin and Ampicillin

E. Type of Action

1. Bacteriostatic

   e.g., Sulphonamides, Chloramphenicol, Erythromycin, Trimethoprim and Clindamycin

2. Bactericidal

   e.g., Penicillin G, Cephalexin, Streptomycin, Vancomycin, Bacitracin, and Potentiated Sulphonamides

F. Source

1. Natural and semi-synthetic

   a. Fungi

      e.g., Penicilin G, Griseofulvin and Cephalexin

   b. Actinomycetes

      e.g., Streptomycin, Tetracycline, Erythromycin and Chloramphenicol.

   c. Bacteria

      e.g., Polymyxin B, Colistin and Bacitracin

2. Synthetic

   e.g., Sulphonamides, Trimethoprim, Quinolones, Nitrofurans and Nitroimidazoles

CHOICE OF AN ANTIMICROBIAL AGENT

When an antimicrobial agent is indicated, its choice should depend on the peculiarities of infecting organism, drug and patient. The aim of chemotherapy should be to produce a selective action on infecting organism for a desirable period with least effect on the host animal.

1. Organism related factors

1. **Target organism:** Identification of target organism is the first step towards selection of an antimicrobial agent. Identification of causative organism can be made from historical data and experience or based on organism identified by culture at the site of infection. A rapid assessment of the nature of organism can be sometimes made on the basis of differential stains such as Gram- stain, but it is often required to culture the infective organism. Newer methods that use molecular biological techniques like Polymerase Chain Reaction (PCR) to identify microorganisms are becoming important

2. **Sensitivity pattern:** In some cases, bacterial sensitivity becomes important to select right antimicrobial drug. This is useful when there is wide variation in the susceptibility of different strains of the same bacterial species to antimicrobial agents.

2. Drug related factors

1. **Spectrum of activity:** For definitive therapy, when target organism has been known and its sensitivity determined, a narrow-spectrum antimicrobial that selectivily affects the concerned organism is preferred. If bacteriological services are not available or treatment cannot be delayed as in serious life-threatening infections, empirical antimicrobial therapy (use of antimicrobial drugs in the absence of an etiological diagnosis) to cover all likely organisms with broad spectrum drug may be used. Polymicrobial infection may also require broad spectrum antimicrobial agent or combination of two or more agents.

2. **Type of activity:** Selection of a bacteriostatic or bactericidal drug depends on certain conditions. A bactericidal antimicrobial agent may be preferred

over bacteriostatic drug in the treatment of life threatening diseases. As a bacteriostatic drug inhibits only multiplication of bacteria and relies upon host defences to clear the infection, it should be used only when host immunity is strong and body systems are functioning normally.

3. **Pharmacokinetic profile:** The pharmacokinetic profile determines the effective concentration of drug at the site of infection for adequate length of time. Antimicrobials such as chloramphenicol and macrolides are extensively metabolized. So they are not used to treat urinary tract infections for which drugs undergoing rapid renal excretion and accumulation in urine is preferred (e.g., Nalidixic acid). The Fluoroquinolones have excellent tissue penetration and attain high concentrations in body tissues.

4. **Route of administration:** The availability of drug penetrations determines the drug selection in many instances because not all drugs are available for administration by all routes. For critically ill patients, parenteral route, in particular intravenous route is preferred. Oral administration of antimicrobial agent is preferred for long-term administration, and when the target is GI infection, Aminoglycosides, Penicilin G, many Cepahalosporins and Vancomycin have to be given by injection only.

5. **Relative toxicity:** Antimicrobial drugs vary considerably in their toxicity potential. Penicillins are among the least toxic of all antimicrobial drugs, so they should be preferred over aminoglycosides and chloramphenicol. However, penicillins are contraindicated in animals that are hypersensitive to the group.

6. **Antimicrobial policy:** Selection of AMA also depends on the predetermined policy so that development of bacterial resistance to antimicrobial drugs may be minimized. Changes in resistance pattern in a particular area should be monitored and antimicrobial therapy altered accordingly. Certain AMAs like fluoroquinolones are not recommended in food animals due to apprehension of developing bacterial resistance in humans.

7. **Cost of therapy:** Often several drugs may show similar efficacy in treating an infection, but vary widely in cost. In such cases, a less costly drug should be preferred.

3. Host factors

1. **Host defence mechanisms:** Elimination of infecting organism depends on the integrity of host-defence mechanisms. Deficient immune capabilities can modify the effectiveness of many antimicrobial drugs, especially those with only bacteriostatic actions. Higher than usual doses of bactericidal agents and longer treatment are required to eliminate infecting organisms in immunocompromised patients.

2. **Pathological conditions:** Pathological conditions like renal insufficiency, hepatic dysfunction alter response and toxic potential of some antimicrobial agents.

3. **Local disorders**

    a. **Pus:** Presence of pus, tissue debris and body secretions decreases efficacy of most AMAs, especially Sulphonamides and Aminoglycosides

    b. **Hematomas:** It favors bacterial growth and impair activity of certain Antimicrobials, Penicillins, Cephalosporins and Tetracyclines bind to degraded hemoglobin in the hematomas

    c. **pH:** The pH of the body fluid determines the efficacy of certain drugs. For example, aminoglycosides are more active in alkaline urine and penicillins are inactivated in acidic pH.

4. **Age:** Renal and hepatic elimination processes are often poorly developed in newborns making neonates particularly susceptible to toxic effects of certain drugs like chloramphenicol and sulphonamides. Tetracyclines are contraindicated in young animals because they accumulate in developing bone and teeth and discolour and weaken them.

5. **Species:** Species differ in their ability to eliminate antimicrobials or are susceptible to adverse effects of drugs.

6. **Brooders:** Antimicrobials should be avoided in brood animals except when essentially required.

7. **Genetic factors:** Certain genetic abnormalities must be considered while choosing AMA. For example, Sulphonamides and Chloramphenicol may produce acute haemolysis in animals with glucose-6-phosphate deficiency.

Adverse reactions to AMAs

The antimicrobial therapy is based on the principle of selective toxicity of a drug for the invading organism rather than the mammalian cells. However, several antimicrobials produce adverse effects in the host, which may be of following types:

1. **Direct tissue toxicity:** Several AMAs cause toxicity due to their adverse effects on cellular processes in the body tissues. These effects mostly occur at high dosages / prolonged use and may include nephrotoxicity (e.g., aminoglycosides), hepatotoxicity (e.g., tetracyclines) and neurotoxicity (e.g., neomycin, streptomycin).

2. **Hypersensitivity reactions:** Hypersensitivity reactions to AMAs or their metabolites may occur with practically all types of AMAs. The hypersensitivity reactions are unpredictable and unrelated to dose and may be immediate or delayed type. Hypersensitivity reactions are more common with some groups of drugs including penicillins, cephalosporins.

3. **Superinfection:** Superinfection refers to the appearance of a new infection as a result of indiscriminate use of AMAs. This primarily occurs due to elimination of normal bacterial flora. Particularly in GIT by AMAs as a result of which certain opportunistic resistant bacterial strains dominate. It is common with the broad spectrum antimicrobials like tetracyclines and occurs frequently with immunocompromised animals.

4. **Nutritional deficiencies:** Certain deficiencies have been associated with chronic use of AMAs. Prolonged use of AMAs generally alters the normal GI flora involved in the synthesis of Vitamin K and some members of Vitamin B.

Failure of antimicrobial therapy

Antimicrobials may fail to cure an infection or there may be relapses, even after the complete chemotherapy course. Failure of chemotherapy may occur due to one or more of the following reasons.

1. Improper diagnosis of disease e.g., viral and not bacterial infection

2. Improper selection of antimicrobial drug e.g., causative organisms are not sensitive to drug.

3. Development of antimicrobial resistance

4. Infection is caused by multiple organisms (mixed infection) and the antimicrobial drug is not effective against all pathogens

5. Antimicrobial penetration into the site of infection is not proper due to the presence of pus, tissue debris etc.

6. Impaired host-defence mechanism is not successful in eradicating static bacteria

7. Selection of route of AMA administration is not proper.

8. Duration of antimicrobial treatment is inadequate

9. Interaction of AMA with concomitantly administered drugs

10. Antimicrobial treatment began too late

11. Use of expired AMA

Guidelines for successful antimicrobial therapy

For the successful antimicrobial therapy and to minimize the emergence of bacterial resistance, following guidelines/policy should be adopted

1. For definitive therapy, use narrow spectrum antimicrobial drug. Keep the broad-spectrum drugs reserved for situations where they are specifically indicated or causative agent/sensitivity is not known.

2. Prefer bactericidal over a bacteriostatic drug

3. Use less toxic AMA than potentially more toxic agent

4. Prefer an AMA that require less frequent administration than that which is given after short intervals

5. For less severe infections, prefer an oral AMA but for severe infections use a parenteral antimicrobial drug

6. Always use AMA in proper dose and for proper duration of time

7. Select an antimicrobial only when the indications are clear

8. Avoid overuse of newer agents when already available agents are effective

9. Do not use antimicrobials to treat untreatable infections

3

DRUG DISCOVERY AND DEVELOPMENT

Sources of drugs

Drugs are derived from various sources and can be classified as natural, synthetic (created artificially), semi-synthetic (containing both natural and synthetic components), and synthesized (created artificially but in imitation of naturally occurring substances).

Natural sources of drugs

Some drugs are naturally occurring biological products and can be made or taken from single-celled organisms, plants, animals, minerals, and humans. Many herbal products come from natural sources, and two life-saving drugs, insulin and penicillin, are also derived from nature. Other examples of modern-day drugs from natural sources include the antibiotic streptomycin produced from cultures of the bacterium *Streptomyces griseus*.

Laboratory sources of drugs

In the modern era, many naturally occurring substances have been combined with other ingredients in a laboratory setting to produce synthetic, semi-synthetic, and synthesized drugs.

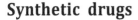

Synthetic drugs

A synthetic drug has been created from a series of chemical reactions to produce a specific pharmacologic effect. Phenobarbital – a barbiturate prescribed for seizure, nerve, or headache disorders – is an example of a synthetic drug. Another example would be a sulfa antibiotic. Both phenobarbital and sulfa are considered synthetic drugs because these substances do not exist in nature.

Semisynthetic drugs

A semisynthetic drug is a natural drug that has been modified chemically in the laboratory to do one or more of the following actions: (1) improve the efficacy of the natural product (2) reduce its side effects (3) overcome developing bacterial resistance or (4) broaden the spectrum of bacteria against which a product can be effective. Many current antibiotics such as Amoxicillin/clavulanate, Azithromycin, and Ampicillin are modifications of existing natural drugs. These antibiotics are more effective against different strains of bacteria or bacteria that have developed resistance to the natural products.

Synthesized drugs

A synthesized drug is created artificially in the laboratory but in imitation of a natural drug. Epinephrine hydrochloride is an example of a synthesized drug. Synthesized drugs have also found their way into the illegal drug market. Known as "designer drugs," these illegal drugs are produced in home chemistry laboratories by individuals who skirt drug control laws by modifying the chemical structure of existing drugs. The new drugs they create offer pharmacologic effects similar to those of their drug counterparts.

Drug nomenclature

Prior to understanding the drug development process, allied health professionals must be familiar with drug nomenclature, or a naming system used to classify drugs. Drug names are assigned according to principles of consistency, logic, and safety in using the name for prescribing, ordering, dispensing, and administering a drug. Every drug has three names: a chemical name, a generic name, and a brand name or trade name.

Chemical name

The chemical name describes the chemical makeup of a drug based on its molecular structure and traditional chemical nomenclature. The chemical name of a drug is typically long and difficult to pronounce, such as N-acetyl-p-aminophenol. This name is used early in a drug's development by the pharmaceutical chemist. Later in the development process, the chemical name will be assigned a generic name the healthcare agency.

Generic name

The generic name is a shorter name that identifies the drug without regard to what company is manufacturing and marketing the drug. Also referred to as a USAN (United Status Adopted Name) or the non-proprietary name, the generic name is often a shortened version of the chemical name and always begins with a lowercase letter. An example of a generic name is acetaminophen (derived from the chemical name N-acetyl-p-aminophenol).

In the United States, the United States Adopted Names Council (USANC) assigns the generic names of drugs. The USANC is sponsored by the American Medical Association, the US Pharmacopeial Convention, and the American Pharmacists Association. The USANC selects simple, unique names based on a chemical or pharmacologic relationship.

Brand name

The brand name or trade name is the name under which a manufacturer markets a drug. This name is chosen by the manufacturer and is exclusive to the company. For that reason, a brand name has a registered mark (®) after the name, indicating that the name has been registered with a national trademark office and is the property of the drug manufacturer. An example of a brand name is Tylenol.

At first, a drug is sold only under its brand name. Once a drug's patent has expired, other manufacturers can make and sell the same drug under its generic name. The same generic drug can be manufactured by different drug companies, each of which can give the drug a different brand name. For example, Prinivil and Zestril are two brand names under which the generic drug Lisinopril is marketed by pharmaceutical companies.

Drug development process

There are more than 181,200 ongoing clinical studies throughout the United States and around the globe as per the U.S. National Institutes of Health. However, since the Food and Drug Administration (FDA) came into being in 1938 just 1,453 drugs have been approved through the end of 2013, according to the Regulatory Affairs Professionals Society. In other words, the success rate of experimental drugs making it to pharmacy shelves or your medicine cabinet is pretty low. According to Medicine.net just five out of every 5,000 preclinical drugs will see the light of day being tested on humans. Further, its statistics show that just one of those five will be approved by the FDA.

How exactly is a drug developed?

Step 1: Drug discovery and development

a. **Drug discovery and target validation:** The first step in the drug development process involves discovery work. Discovery begins with target identification – choosing a biochemical mechanism involved in a disease condition. Drug candidates are tested for their interaction with the drug target. Typically, researchers discover drugs through

- New insights into a disease process that allow researchers to design a product to stop or reverse the effects of the disease
- Many tests of molecular compounds to find possible beneficial effects against any of a large number of diseases

- Existing treatments that have unanticipated effects
- New technologies, such as those that provide new ways to target medical products to specific sites within the body or to manipulate genetic material

At this stage in the process, thousands of compounds are tested for their interaction with the target. Each potential drug (hits) candidate is subjected to screening process. Once interaction with the target drug is confirmed, target may be validated by checking for activity *vs* disease condition. After careful review, only small number of compounds looks promising and calls for further study.

b. **Development:** Once researchers identify a promising compound for development, they conduct experiments to gather information on:

- How it is absorbed, distributed, metabolized and excreted
- Its potential benefits and mechanisms of action
- The best dosage
- The best way to give the drug (such as by mouth or injection)
- Side effects (often referred to as toxicity)
- How it affects different groups of people
- How it interacts with other drugs and treatments
- Its effectiveness as compared with similar drugs

Step 2: Preclinical Research

The next step in the drug development process is preclinical testing, which in itself is divided into two subcomponents: *in vitro* and *in vivo* testing. *In vitro* testing examines the drug molecule's interactions in test tubes and within the lab setting. *In vivo* testing involves testing the drug molecules on animal models. FDA requires researches to use Good Laboratory Practices (GLP), defined in medical product development regulations, for preclinical laboratory studies. These regulations set the minimum basic requirements for:

- Safety conduct
- Personnel

- Facilities
- Equipment
- Written protocols
- Operating procedures
- Study reports
- System of quality assurance oversight for each study to help assure safety of FDA regulated product

Usually, preclinical studies are not very large. However, these studies must provide detailed information on dosing and toxicity levels. After preclinical testing, researchers review their findings and decide whether the drug should be tested in people.

Step 3: Clinical research

"Clinical research" refers to studies, or trials, that are done in people. As the developers design the clinical study, they will consider what they want to accomplish for each of the different clinical research phases and begin the Investigational New Drug Process (IND), a process they must go through before clinical research begins.

Clinical research phase studies

Once the animal studies are completed, 3 phases of clinical trials are to be performed. Protocol for testing should be developed by researchers and get it approved by FDA. The protocol describes, what type of subjects may participate, schedule of the test, dosage, length.

Phase 1 (to study the safety and dosage)

During Phase 1 studies, researchers test a new drug in normal volunteers (healthy individuals). In most cases, 20 to 80 healthy individuals with the disease /

condition participate in Phase 1. However, if a new drug is intended for use in cancer patients, researchers conduct Phase 1 studies in patients with that type of cancer. Phase 1 studies are closely monitored and gather information about how a drug interacts with the human body. Researchers adjust dosing schemes based on animal data to find out how much of a drug the body can tolerate and what its acute side effects are. As a Phase 1 trial continues, researchers answer research questions related to how it works in the body, the side effects associated with increased dosage, and early information about how effective it is to determine how best to administer the drug to limit risks and maximize possible benefits. This is important to the design of Phase 2 studies. Approximately 70% of the drugs move to next phase.

Phase 2 (to study the efficacy and side effects)

In Phase 2 studies, researchers administer the drug to a group of patients with the disease or condition for which the drug is being developed. Typically involving a few hundred patients, these studies aren't large enough to show whether the drug will be beneficial. Instead, Phase 2 studies provide researchers with additional safety data. Researchers use these data to refine research questions, develop research methods, and design new Phase 3 research protocols. Approximately 33% of drugs move to the next phase.

Phase 3 (to study the efficacy monitoring of adverse reactions)

Researchers design Phase 3 studies to demonstrate whether or not a product offers a treatment benefit to a specific population. Sometimes known as pivotal studies, these studies involve 300 to 3,000 participants. Phase 3 studies provide most of the safety data. In previous studies, it is possible that less common side effects might have gone undetected. Because these studies are larger and longer in duration, the results are more likely to show long-term or rare side effects. Approximately 25-30% of drugs move to the next phase.

Phase 4 (to study the safety and efficacy)

Phase 4 trials are carried out once the drug or device has been approved by FDA during the Post-Market Safety Monitoring.

The investigational new drug process

Drug developers, or sponsors, must submit an Investigational New Drug (IND) application to FDA before beginning the clinical research. In the IND application, developers must include:

- Animal study data and toxicity (side effects that cause great harm) data
- Manufacturing information
- Clinical protocols (study plans) for studies to be conducted
- Data from any prior human research
- Information about the investigator

Asking for FDA assistance

Drug developers are free to ask for help from FDA at any point in the drug development process, including:

- Pre-IND application, to review FDA guidance documents and get answers to questions that may help enhance their research
- After Phase 2, to obtain guidance on the design of large Phase 3 studies
- Any time during the process, to obtain an assessment of the IND application

Even though FDA offers extensive technical assistance, drug developers are not required to take FDA's suggestions. As long as clinical trials are thoughtfully designed, reflect what developers know about a product, safeguard participants, and otherwise meet Federal standards, FDA allows wide latitude in clinical trial design.

FDA IND review team

The review team consists of a group of specialists in different scientific fields. Each member has different responsibilities.

- **Project Manager:** Coordinates the team's activities throughout the review process, and is the primary contact for the sponsor.
- **Medical Officer:** Reviews all clinical study information and data before, during, and after the trial is complete.
- **Statistician:** Interprets clinical trial designs and data, and works closely with the medical officer to evaluate protocols and safety and efficacy data.
- **Pharmacologist:** Reviews preclinical studies.
- **Pharmakineticist:** Focuses on the drug's absorption, distribution, metabolism and excretion processes. Interprets blood-level data at different time intervals from clinical trials, as a way to assess drug dosages and administration schedules.
- **Chemist:** Evaluates a drug's chemical compounds. Analyzes how a drug was made and its stability, quality control, continuity, the presence of impurities, etc.
- **Microbiologist:** Reviews the data submitted, if the product is an antimicrobial product, to assess response across different classes of microbes.

Approval

The FDA review team has 30 days to review the original IND submission. The process protects volunteers who participate in clinical trials from unreasonable and significant risk in clinical trials. FDA responds to IND applications in one of two ways:

- Approval to begin clinical trials.
- Clinical hold to delay or stop the investigation. FDA can place a clinical hold for specific reasons, including:
  o Participants are exposed to unreasonable or significant risk.
  o Investigators are not qualified.

o   Materials for the volunteer participants are misleading.

o   The IND application does not include enough information about the trial's risks.

A clinical hold is rare; instead, FDA often provides comments intended to improve the quality of a clinical trial. In most cases, if FDA is satisfied that the trial meets Federal standards, the applicant is allowed to proceed with the proposed study.

The developer is responsible for informing the review team about new protocols, as well as serious side effects seen during the trial. This information ensures that the team can monitor the trials carefully for signs of any problems. After the trial ends, researchers must submit study reports.

This process continues until the developer decides to end clinical trials or files a marketing application. Before filing a marketing application, a developer must have adequate data from two large, controlled clinical trials.

4

COMBINATION THERAPY

Combination therapy (or polytherapy) is a broad term for the use of multiple medications or therapies, in order to fight the same condition. It is widely known that many strains of mostly Gram-negative bacteria have become and are still becoming increasingly resistant to current antibiotic drugs. With the development of new antibacterial drugs slowing down due to a number of reasons, scientists are looking to study combination therapy for use against bacteria which includes use of two or more drugs together to restore or increase the efficacy of both drugs against the bacterial pathogen that is resistant to ordinary antibiotics.

Single Agent Chemotherapy

Single most specific drug is preferable if the infecting bacteria have been identified. Administration of a single drug with a narrow spectrum of action is desirable because:

- Alteration of normal flora is minimized (which in turn reduces the likelihood of overgrowth of resistant bacteria)

- Reduces toxicity which may be associated with multiple drug regimens

- Reduces cost

Combination chemotherapy

Interference between two antibiotics can only be expected if their mode of action is different. Antibiotics may masquerade under different names, and may even be obtained from widely different sources, and yet be the same as regards their antibacterial action even though their pharmacological properties may be different. Combination of such related drugs have the same effect as increasing the effective concentration of either alone, and is therefore of little interest. The term ' additive' may be applied to such a combination. For true interaction, the drugs must have a different point of attack in the bacterial metabolism. If two vital processes are blocked at the same time, the result may be either a greater or a lesser effect than with either of the two drugs alone. If two antibiotics, each with a bactericidal effect, interfere even in subminimal concentration with two distinct and essential vital processes, death of the bacterial cell may result although neither of the two drugs was present in an amount necessary to cause death by itself. The opposite effect may also occur. Bactericidal drugs are generally most active on growing, rapidly metabolizing cell populations. If a bacteriostatic drug slows down the rate of turnover of the microbe, a bactericidal drug may be unable to exert its full deadly blow.

Although single-agent antimicrobial therapy is generally preferred, a combination of 2 or more antimicrobial agents is recommended in a some cases.

1. When agents exhibit synergistic activity against a microorganism: Synergy between antimicrobial agents means that, when studied *in vitro*, the combined effect of the agents is greater than the sum of their independent activities when measured separately.

2. When critically ill animals / patients require empiric therapy before microbiological etiology and/or antimicrobial susceptibility can be determined, Combination therapy is used in this setting to ensure that at least one of the administered antimicrobial agents will be active against the suspected organism

3. To extend the antimicrobial spectrum beyond that achieved by use of a single agent for treatment of polymicrobial infections: When infections are thought to be caused by more than one organism, a combination regimen may be preferred because it would extend the antimicrobial spectrum beyond that achieved by a single agent.

4. To prevent the emergence of resistance: The emergence of resistant mutants in a bacterial population is generally the result of selective pressure from antimicrobial therapy. Provided that the mechanisms of resistance to 2 antimicrobial agents are different, the chance of a mutant strain being resistant to both antimicrobial agents is much lower than the chance of it being resistant to either one. In other words, use of combination therapy would provide a better chance that at least one drug will be effective, thereby preventing the resistant mutant population from emerging as the dominant strain and causing therapeutic failure.

The selection of antimicrobials for combination therapy should be based on

- Mechanism of action that are different
- Spectrum of activity that are complimentary

Jawetz's law on antimicrobial combination

- Bactericidal + bactericidal: may be synergistic or additive (indifferent).
- Bacteriostatic + bacteriostatic: usually additive.
- Bacteriostatic + bactericidal: may be antagonistic or indifferent

In vitro results of combination therapy

1. Synergism can be defined as the positive interaction of two or more agents so that their combined effect is significantly greater than the expected sum of their individual effects (1 + 1 = 3). For example, combination of penicillin and streptomycin produces lethal effect on Streptococci. Because penicillin enhances the penetration of aminoglycosides by causing bacterial cell wall damage.

2. Antagonism is a negative interaction between two drugs; the combined effect of their combination is significantly less than the sum of the respective effect when tested separately (1+1=1). For example, combination of tetracycline and penicillin produces reduced antibacterial activity as former drug inhibit bacterial growth and multiplication that is essentially required for cell wall damaging effect produced by β-lactam antibiotics.

3. Additivity assumes that the result observed with more than one drug should be the sum of the separate effects of the drug being tested if those drugs do not interact with one another (1+1=2).

5

BACTERIAL RESISTANCE TO ANTIMICROBIALS

B acterial resistance to AMAs refers to unresponsiveness of a microorganism to an antimicrobial drug even at the maximal level that is tolerated by the host. It is of two types – natural and acquired

1. Natural resistance

When an organism is inherently or genetically resistant to an AMA, the resistance is called natural resistance. Natural resistance to drugs in an organism occurs due to many factors including lack of penetration to drug into bacterial cell, absence of metabolic pathway or target site affected by drug or rapid inactivation of drug in the bacterial cell. Natural resistance is generally a characteristic of a group or a genus of bacteria and it is well known. For example, Gram-negative organisms are resistant to penicillin G and vancomycin.

2. Acquired resistance

When an organism becomes resistant to an AMA to whom it was previously sensitive, the resistance is called acquired resistance. The acquired resistance develops over a period of time and initially is not known. Unlike, natural resistance, acquired resistance can happen with any microbe and is a great threat to the antimicrobial therapy. Acquired resistance in the microorganism has been ascribed

in the widespread and inappropriate use of antimicrobial drugs. The antimicrobial drugs do not induce resistance but generally help in selection of bacterial strains that are inherently resistant.

MECHANISMS OF ACQUIRED RESISTANCE

Transmission

Acquired resistance may develop by mutation and gene transfer.

Mutation

Mutation is a stable and heritable gene change that occurs spontaneously and randomly among the microorganisms. Mutation occurs due to insertion, deletion and substitution of one or more nucleotides in the genome which may be present on chromosomes or on extra chromosomal materials called plasmids. The resulting alteration in the genome may persist, be corrected or be lethal to the cell. If the mutation remains unrepaired and the cell survives, the altered genome is transmitted to daughter cells on cell replication, thus producing resistant strains or mutants.

Mutation is usually not induced by drugs, but drugs may act as selective force that favour emergence of mutants. Mutation occurs normally once in 10^8 cell divisions, so any sensitive population of microbes at a given time may contain few mutant cells. If the mutants are resistant to MAAs, they are selectively preserved and get chance to proliferate when the antimicrobial agent eliminates the sensitive cells in time, the resistant strains predominates and replaces the sensitive strain. Mutational acquisition of resistance in microorganisms is generally less important because in most cases, the acquired resistance is accompanied by other changes that render the organism less viable with decreased virulence. Therefore, such mutants are more likely to be destroyed by other drugs or are removed by host system.

Acquired resistance by mutation occur either in a single step or in a series of steps:

1. **Single step mutation:** In this type, bacterial resistance develops in a single step due to mutation occurring in a powerful gene. It emerges rapidly and confers high degree of resistance.

2. **Multistep mutation:** In this type, bacterial resistance develops in multiple steps due to mutation occurring in a number of different genes. Multiple mutation develops slowly and gradually and confers a slight resistance.

Gene transfer

Resistance by gene transfer (infectious resistance) develops by transfer of genetic material coding for resistance (R factor) from a resistant microorganism to a susceptible organism. The origin of R factor is not clearly known, but their development is possibly not related to antimicrobial therapy. The resistance properties are usually (not exclusively) encoded in the extra-chromosomal DNA, the plasmid. Resistance genes can be transferred between bacteria through a number of ways including transduction, transformation and conjugation

1. **Transduction:** It is the transfer of gene carrying resistance (R factor) by the intervention of a bacteriophage (virus that infects bacteria). The bacteriophage usually utilizes the bacterial machinery for multiplication and during the process incorporate R factor into the genetic material of viral progeny. When the progeny viral particles infect other bacteria, the genetic information encoding resistance is passed to the bacteria. The newly infected bacteria may become resistant to the antimicrobial drug and contribute to pass on the resistance to their progeny. Transduction is a less common method by which organisms acquire resistance. Transduction is involved in the transfer of resistance in *Staphylococcus aureus* and streptococci to penicillin by inducing penicillinase production.

2. **Transformation:** In this process of gene transfer, a sensitive bacterium transforms to resistant one by acquiring the free resistance carrying DNA material from the medium/environment. Such free DNAs are secreted or released by the resistant bacteria, usually of the same strain or closely related strains into the medium after their lysis. Acquisition of resistance by transformation is relatively infrequent and is of little clinical significance.

3. **Conjugation:** It is a type of reproductive process in which resistant genes (R factors) transferred from one bacterium to another by direct contact through a pilus or bridge. The formation of pilus is coded by Resistance Transfer Factor (RTF) on a plasmid. Thus genes carrying resistance (R factor) and RTF together are involved in transferring resistance *via* conjugation.

Conjugation is now recognized as the most important mechanism for spread of antimicrobial resistance. In the transfer of resistance genes, the donor and recipient strain may belong to different species, different genera, (among Gram-ve and between Gram-ve and Gram+ve organisms) and pathogenic and non-pathogenic organisms. A single gene transfer of resistant genetic material that codes for resistance to multiple drugs from a bacterial donor can result in antimicrobial resistance against several drugs in the recipient bacterial cell.

Biochemical mechanism of resistance

The biochemical mechanisms of bacterial resistance vary and involve drug penetration, binding sites, metabolic pathways or drug inactivation enzymes

1. **Alteration in drug penetration:** Decreased penetrability of an antimicrobial agent into the bacterial cell may occur due to either changes in permeability of cell wall or alterations in transport systems. Resistance in Gram-ve bacteria may occur due to decreased porin sizes or numbers, which normally allow intracellular accumulation of hydrophilic antimicrobials like aminoglycosides. Some resistant bacteria cause rapid efflux of antimicrobial drug (e.g., tetracyclines) *via* energy dependent carrier processes. In such cases, the reduced concentration of an antimicrobial in bacteria fails to produce antimicrobial action because it is unable to gain access to the site of action.

2. **Alteration in binding sites:** Molecular alterations in target sites in an organism can confer resistance because antimicrobials fail to bind to their specific sites and thus cannot perform their action. For example, alterations in proteins of the 30S ribosomal subunit impart resistance to aminoglycosides and changes in structure of 50 S ribosomal subunits impart resistance in penicillin binding proteins can confer resistance against penicillins. The point mutation of DNA gyrase may lead to decreased affinity for quinolones.

3. **Alterations in metabolic pathways:** Alteration in the metabolic pathways may impart resistance by bypassing the reactions inhibited by the AMA. For example, some sulphonamides resistant bacteria do not require intracellular para amino benzoic acid (PABA) unlike mammalian cells that utilize preformed folic acid.

4. **Drug inactivating enzymes:** The increased production of drug inactivating enzymes by certain microbes also confers resistance on microorganism. These drug metabolizing enzymes are either inducible or constitutive. For example, beta lactmases destroy many penicillins and cephalosporins by cleaving the beta lactam ring. Similarly, aminoglycoside are inactivated by certain transferases.

Cross resistance

Cross resistance is a type of acquired resistance in which bacteria resistant to one AMA also become resistant to another AMA without having exposed to the latter. It is commonest among the chemical drugs, but may also seen in chemically unrelated drugs that share the same mechanism of action or attachment. Cross resistance among antimicrobials is of two types.

1. **Complete cross resistance / two way cross resistance:** In this type of cross resistance, bacteria resistant to one antimicrobial is also resistant to a second drug and *vice versa*. For example, cross resistance between Neomycin and kanamycin and erythromycin and Oleandomycin is complete

2. **Partial cross resistance / one way cross resistance:** In this type of cross resistance, bacteria resistant to one AMA is also resistant to a second drug but resistant to second AMA does not lead to resistant to first AMA. For example, resistance to Gentamicin leads to resistance to Kanamycin and Streptomycin, but Kanamycin and Streptomycin does not necessarily extend to Gentamicin.

6

QUINOLONES: DNA SYNTHESIS
INHIBITORS

As we know cell division is essential for an organism to grow, but, when a cell divides, it must replicate the DNA in its genome so that the two daughter cells have the same genetic information as their parent. The double-stranded structure of DNA provides a simple mechanism for DNA replication. Usually DNA is coiled inside the cell, because DNA is very large ($3x10^9$ bps) supercoiling is important in a number of biological processes, such as compacting DNA and by regulating access to the genetic code. DNA supercoiling strongly affects DNA metabolism and possibly gene expression. Supercoiling (coiling of the coiled DNA) of DNA reduces the space and allows for DNA to be packaged. The DNA molecule present within the bacterial cell is negatively supercoiled all the time. During replication, helicase starts to separate the two strands. As helicase moves in a direction, it is creating enormous pressure on other regions of DNA, which was negatively supercoiled. Due to the enormous tension applied, it will start to have a very complicated structure. It will eventually shear. But the cell needs to make this DNA intact at its place. For that, it needs to uncoil a little bit. To counteract the force, the DNA needs to be positively coiled. This can happen with the help of certain enzymes such as topoisomerases. They can change DNA topology to facilitate functions such as DNA replication or transcription.

Topoisomerases such as DNA gyrase (topoisomerase II) and topoisomerase IV act in concert to maintain an optimum supercoiling state of DNA in the cell.

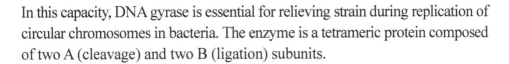

In this capacity, DNA gyrase is essential for relieving strain during replication of circular chromosomes in bacteria. The enzyme is a tetrameric protein composed of two A (cleavage) and two B (ligation) subunits.

Quinolones

Quinolones are a group of synthetic antibacterial agents having a 4-quinolone structure. They are primarily active against gram-negative bacteria, although newer fluorinated agents (fluorinated quinolones) also inhibit selected gram-positive bacteria. They are minimally toxic and are becoming important in veterinary medicine.

History

Nalidixic acid was the first member of quinolones family introduced in 1964. As its usefulness was limited to urinary and GI tract infections, its congener's oxolinic acid and rosoxacin with more potency but limited spectrum were introduced in 1970s. This was followed by second-generation quinolones called fluoroquinolones with extended spectrum and systemic antibacterial effects in 1980s. Since then, large number of fluorinated quinolones have been synthesized and found place in human and/or veterinary practices.

Chemistry and structure-activity relationship

Currently available quinolones contain basic structure of 4-quinolones (short for 4-oxo-1,4-hydroquinoline) with a carboxylic acid moiety in the position 3.

Various modifications on the basic ring structure have produced compounds with differing physical, chemical, pharmacokinetic and antimicrobial properties. On the basis of structure, following generalisations can be made:

a) The carboxyl group at position 3 and ketone group at positon 4 are essential for antibacterial activity.

b) Substitution at position 6 with a fluorine moiety markedly enhances activity against gram-negative and gram-positive bacteria. All fluorinated 4-Quinolones are called fluoroquinolones.

c) Addition of a piperazine ring at position 7 on fluoroquinolones significantly increases tissue and bacterial penetration and improves spectrum of activity against Pseudomonas (e.g., Ciprofloxacin, Enrofloxacin).

d) Substitution with an oxygen atom at position 8 improves activity against gram-positive and anaerobic organisms without affecting the bactericidal profile.

e) Change to a carbon from a nitrogen at position 8 decreases some adverse CNS effects and increases activity against staphylococci.

Mechanism of action

Quinolones are bactericidal drugs that inhibit replication of bacterial DNA by interfering with the action of DNA gyrase (topoisomerase). Quinolones enter susceptible microorganisms by the passive diffusion through water-filled channels in the outer membrane. Once inside the bacterial cell, quinolones target and DNA gyrase that is an enzyme which controls the supercoiling of DNA and converts released covalently closed circular DNA to a super helical form by energy dependent strand breakage and resealing. Super coiling is a process of coiling of double stranded DNA molecule on itself so that a DNA up to 1.3mm length can be tightly and compactly packed inside the bacterial cell. The supercoiling requires first nicking of the double stranded DNA and then resealing it after the super twist (passage of double stranded DNA though the nick). This is carried out by DNA gyrase in an energy dependent reaction. Bacterial DNA gyrase is composed of two A and two B subunits. The A subunits carry out the strand-cutting function, whereas the B subunits cause the ATP hydrolysis necessary for gyrase supercoiling. The A subunits are the primary site of action of quinolones. When DNA gyrase is inhibited by quinolones, a reduction in the supercoiling occurs with consequent disruption of the DNA replication. The exposed nicks or cuts in the DNA that are not resealed also lead to degradation of DNA into small fragments by the action of exonucleases.

Bacterial DNA gyrase is susceptible to fluoroquinolones at concentrations of 0.1 to 10 μg/ml and the susceptible microorganisms lose viability within few minutes of exposure to the antibacterial. Mammalian topoisomerases also perform similar nicking, but the enzymes are fundamentally different from bacterial gyrases and are not susceptible to quinolones. Further, affinity of mammalian topoisomerase for quinolones is very low (0.001) than that of bacterial DNA gyrase. An unusual feature of the group is that at lower concentration (below MIC values) and at higher concentrations than the optimal range, the fluorquinolones are less active against susceptible bacteria. This biphasic pattern is probably caused by depression of RNA synthesis at high concentrations.

Antimicrobial spectrum

First-generation quinolones (e.g., nalidixic acid) tend to have only a moderately extended gram-negative spectrum. The spectrum of activity of fluoroquinolones might be considered broad with efficacy against a wide range of gram-negative and gram-positive bacteria, *Mycoplasma,* and *Chlamydia*. However, spectrum of activity varies with the type of fluoroquinolone. Common bacteria susceptible to fluoroquinolones in small animals include *E. coli, Salmonella, Campylobacter, Shigella, Brucella, and Vibrio*. They are also effective against *Pseudomonas aeruginosa* and *Staphylococcus*, but *Streptococcus* tends to be resistant to most of the fluoroquinolones. Fluoroquinolones are not effective against anaerobes.

Bacterial resistance

Bacterial resistance develops rapidly to nalidixic acid and some other older quinolones. Resistance to fluoroquinolones develops slowly and is of low grade. Resistance noted so far is either due to chromosomal mutation producing alteration in bacterial DNA gyrase with a decreased affinity for quinolones or due to reduced permeability of bacterial membranes to quinolones. Energy-dependent efflux of drugs out of the bacteria is also possible.

Pharmacokinetics

Pharmacokinetics of quinolones varies with type of drug and species.

Absorption: In general, quinolones have a good rate and extent of absorption after oral administration. Bioavailability is often>80% for most quinolones.

Magnesium and calcium ions decrease absorption of quinolones after oral administration; absorption from IM injection is rapid.

Distribution: The quinolones with few exceptions distribute well into all body tissues and fluids. The degree of plasma protein binding is extremely variable from ~10% for norfloxacin to >90% for nalidixic acid.

Biotransformation and excretion: The elimination of quinolones depends on the type of agent. Some quinolones are eliminated unchanged (e.g., ofloxacin), some are partially metabolized (e.g., ciprofloxacin), and a few are completely metabolized (e.g., pefloxacin). Metabolites of a few quinolones are active i.e., enrofloxacin undergoes de-ethylation to form ciprofloxacin. Renal excretion is the major route of elimination for most quinolones. Some fluoroquinolones undergo hepatic as well as renal clearance.

Side effects/Adverse effects

Side effects with older quinolones are relatively common, but newer quinolones are generally tolerated well. Fluoroquinolones can generally tolerate well. They can produce crystalluria due to their low solubility in acidic urine. Haemolytic anaemia has been reported in some animals with fluoroquinolones.

Drug interactions

Quinolones have few but important drug interactions. They are potent chelators of Mg^{++} and Ca^{++} and may interfere with their absorption. A synergistic effect of quinolones with beta-lactam, aminoglycoside, clindamycin, or metronidazole antibacterials has been reported.

Classification

Quinolones are classified in a number of ways like on the basis of their chemical structure, antibacterial spectrum, or pattern of evolution (generation). The most appropriate and simple classification is on the basis of their generation.

I. First-generation quinolones/Non-fluorinated quinolones

e.g., Nalidixic Acid, Oxolinic Acid, Pipemidic Acid, Piromidic Acid and Cinoxacin.

II. Second-generation quinolones/Fluorinated quinolones

e.g., Enrofloxacin, Difloxacin, Ciprofloxacin, Orbifloxacin, Danofloxacin, Marbofloxacin, Norfloxacin, Ofloxacin, Pefloxacin, Sarafloxacin, Iomefloxacin and Enoxacin.

I. FIRST-GENERATION QUINOLONES/NON-FLUORINATED QUINOLONES

Nalidixic acid

Nalidixic acid is a non-fluorinated quinolone used primarily as a urinary antiseptic. It is active against gram-negative bacteria, especially *E. coli, Proteus, klebsiella, Enterobacter, and Shigella*. Gram-positive bacteria, *Pseudomonas aeruginosa*, and anaerobes are not susceptible to nalidixic acid.

Nalidixic acid is well absorbed orally (>90%). It is highly plasma protein bound (>95%). It is partially metabolized in liver by hydroxylation to a more potent bactericidal compound 7-hydroxynalidixic acid, which is excreted in the urine along with parent drug. Due to extensive plasma protein binding, the plasma concentration of free drug is inadequate for treatment of systemic infections.

Oxolinic Acid

It is a non-fluorinated first generation quinolone with the same mechanism of action as that of nalidixic acid. It is one of the most extensively researched of all the drugs used in fish medicine. The standard oral recommendation is 10mg/kg/day in freshwater species. In marine species a higher dose rate is required because of the complexing of the drug with divalent cations.

II. SECOND-GENERATION QUINOLONES/ FLUOROQUINOLONES

The second-generation quinolones (fluoroquinolones) contain fluorine at position 6 on the ring structure. These agents have several advantages over the older non-fluorinated quinolones and have become a separate class of antimicrobial drugs. Important advantages include extended spectrum of activity against both gram-negative and gram-positive organisms and mycoplasma, high potency, rapid bactericidal action, systemic effects, lower incidence of adverse effects and administration *via* a variety of routes.

Enrofloxacin

Enrofloxacin is a prototypic veterinary fluorinated quinolone developed exclusively for use in animals. It occurs as a pale yellow, crystalline powder that is slightly soluble in water. Enrofloxacin is related structurally similar to the human-approved drug ciprofloxacin, having an additional ethyl group on the piperazinyl ring. Like other quinolones, inhibition of DNA gyrase in susceptible bacteria is the primary mechanism of action. It has higher *in vitro* activity against gram-positive bacteria than other quinolones. It has been used for the treatment of bacterial kidney disease, BKD (*Renibacterium salmoninarum*) at a dose of 2.5 mg/kg/day for 10 days and 10 mg/kg/day for 10 days for Furunculosis. 10 mg/kg/day has been recommended for the control of *Vibrio anguillarium* infection.

Pharmacokinetics

Enrofloxacin is well absorbed after oral administration with bioavailability of about 80%. The peak plasma concentration is attained within one hour of oral dosing. Absorption of enrofloxacin is nearly complete from IM injection sites. It is distributed throughout the body. It also accumulates in very high concentrations in WBCs. Enrofloxacin is eliminated primarily unchanged in urine, although up to 25% of the drug is metabolized to the active metabolite, ciprofloxacin, which subsequently is metabolized to inactive metabolites.

Sarafloxacin

It is presented as white to yellow free-flowing powder (Sarafin®) which is 100% sarafloxacin hydrochloride, equivalent to 82% sarafloxacin. It is authorized for use to the treatment of furunculosis (*Aeromonas salmonicida*) in Atlantic Salmon in seawater at a dose regimen of 10 mg/kg/day for 5 days. Sarafin may be incorporated in feed prior to pelleting or mixed with feed just prior to feeding. For on-farm mixing it is recommended that product should first be suspended in edible vegetable oil.

Difloxacin

Difloxacin is a fluorinated quinolone marketed as veterinary product. The spectrum of activity of difloxacin is broad including many gram-negative and gram-positive bacilli and cocci. Some strains of *Pseudomonas aeruginosa* and most *Enterococcus* sp. are resistant. Difloxacin is well absorbed from GI tract after oral administration with peak plasma levels occurring at about 3 hours post-dosing. The drug is well distributed to body tissues and fluids and is marginally bound to plasma proteins (20-50%). It is eliminated principally by hepatic metabolism to inactive compound, sarafloxacin by demethylation, but the amount of active metabolite is small. Unlike most fluorquinolones, difloxacin is excreted mainly in faeces (~80%); renal clearance accounts for less than 5% of the total systemic clearance.

Ciprofloxacin

Ciprofloxacin is a broad-spectrum highly potent fluoroquinolone. It is structurally and pharmacologically related to enrofloxacin. The MIC of ciprofloxacin for gram-negative organisms is usually <-0.1 µg/m, while gram-positive bacteria are inhibited at relatively higher concentrations.

Ciprofloxacin is well absorbed after oral administration in most species. However, the bioavailability of ciprofloxacin after oral administration is low to moderate. Plasma protein binding varies from 25% to 70%. Ciprofloxacin is well distributed to almost all body tissues and fluids. Ciprofloxacin is partly metabolized in liver and excreted primarily in urine by glomerular filtration and tubular secretion.

Marbofloxacin

Marbofloxacin is a fluorinated quinolone with considerably long half-life of 10 hours. It has broad-spectrum of activity and is well distributed to tissues. Marbofloxacin is used for the treatment of bacterial infections susceptible to it.

Norfloxacin

Norfloxacin is primarily used in human medicine for treatment of urinary, genital, and GI tracts infections. It is not recommended for systemic infections because it attains lower concentrations in tissues.

Flumequine

Flumequine is a fluorinated quinolone having predominantly gram-negative activity. Its spectrum of activity covers not only Gram-negative but also fungi, protozoa and even some helminthes. It is occasionally indicated in veterinary practice for treatment of enteric infections (colibacillosis and salmonellosis) in animals. It is gradually replacing oxolinic acid in fish medicine because of its more appropriate pharmacokinetic profile.

Ofloxacin

Ofloxacin is another fluorinated quinolone with antibacterial spectrum similar to enrofloxacin and ciprofloxacin. It is relatively lipid soluble, so its oral bioavailability is high and it also attains high plasma concentrations than some commonly used quinolones.

NEWER FLUOROQUINOLONES

The newest generations of fluoroquinolones, referred to as the second generation fluoroquinolones, include Trovafloxacin, Grepafloxacin, Gatifloxacin, Sparfloxacin, Moxifloxacin, and Premafloxacin. These fluoroquinolones with substitutions at the C-8 position have as their advantage of broader spectrum that includes anaerobic bacteria and Gram-positive cocci.

7

B-LACTAMS: CELL WALL SYNTHESIS INHIBITORS

Beta–lactam antibiotics constitute one of the most important and frequently used antimicrobial agents. This group comprises the penicillins, cephalosporins, and carbapenems. Monobactams are the newer addition. All B-lactam antimicrobial agents share many features such as chemical structure, mechanism of action, pharmacological effects and immunological characteristics.

Penicillins

The penicillins are a large group of naturally occurring and semi-synthetic antibiotics. They have a common nucleus, the 6 – aminopenicillanic acid (6-APA), and a common mode of action. Many member of this group are drugs of choice for a large number of infectious diseases in animals.

History

History of penicillins goes back to 1928 when Sir Alexander Fleming, a British bacteriologist, who found that staphylococcal colonies near a mould contaminating one of his cultures undergo lysis. He elaborated that the mould released an antibacterial substance that does not allow the growth of bacteria. Because the mould belonged to the genus *Penicillium*, Fleming named the antibacterial substance as penicillin. A decade later in 1940, penicillin was isolated and

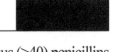

developed as a systemic antibacterial agent. Since then, numerous (>40) penicillins have been discovered and many are being used in clinical practice. Newer compounds are obtained semi-synthetically by chemical manipulations of 6-aminopenicillanic acid.

Chemistry

All penicillins contain the basic structure of a 5 membered thiazolidine ring (A) connected to a beta-lactam ring (B) having a secondary amine group (-NH-) to form 6-aminopenicillanic acid (6-APA), so called the basic nucleus of penicillins. The beta – lactam ring is a four membered ring in which an amide linkage join a carbonyl group and a nitrogen. A side chain (R) is attached to the beta – lactam ring, which mainly determines the type of penicillin. For the general structure, following observations can be made regarding the properties, antibacterial activity and bacterial resistance.

a) The beta – lactam ring is the key structural feature of all beta-lactam antibiotics (e.g., penicillins, cephalosporins, carbapenems). Cleavage of beta – lactam ring destroys antibiotic activity;some resistant bacteria produce beta-lactamase (penicillinase) that cleaves B – lactam ring.

b) The 6-aminopenicillanic acid found in all penicillins is responsible for antibacterial activity and determines the type of penicillin group. Chemical alteration of 6-APA nucleus leads to loss of antibacterial activity.

c) The side chain attached to B-lactam ring determines mainly the individual penicillin characteristics.

d) The carboxyl group attached to thiazolidine ring is the site of salt formation (e.g.,sodium, potassium, procaine). Conversion to salt ester stabilises the penicillins and affects solubility and pharmacokinetics.

e) Cleavage of the amide bond side-chain by amidase yields the 6-APA nucleus which is used in producing semi-synthetic penicillins.

Properties and solubility

The penicillins are poorly soluble weak organic acids that are sensitive to heat, light extremes in pH, heavy metals, strong alcohols and oxidising and reducing agents. A pH of 6-6.5 is optimal for stability of penicillins in aqueous solution with a practical range of 5.5-7.5. A prolonged exposure of penicillins to water promotes hydrolysis. So many penicillins require reconstitution with a diluent just before injection. Some penicillins are rapidly hydrolysed and inactivated by gastric acid and B-lactamase and acylase. Gastric acid hydrolyses the amide side chain and opens the B-lactam ring, whereas B-lactamase presents in some microorganisms cleaves the B-lactam ring producing penicilloic acid derivatives. Aqueous solutions of alkaline sodium salt of sulphonamides also inactivate penicillins. The sodium and potassium salts of natural penicillins enhance water solubility hence are used parenterally. The trihydrate forms of the semi-synthetic penicillins have high water solubility and are usually administered by both parenteral and oral routes.

Classification

Penicillins are broadly divided into following groups depending on their spectrum of antibacterial activity and B-lactamase (penicillinase) sensitivity:

1. Narrow-spectrum penicillins

2. Broad-spectrum penicillins

3. Extended-spectrum penicillins/Anti-pseudomonal penicillins.

4. Potentiated penicillins/Beta-lactamase protected penicillins

Mechanism of action

Bacteria are usually single-celled, except when they exist in colonies. These ancestral cells reproduce by means of binary fission, duplicating their genetic material and then essentially splitting to form two daughter cells identical to the parent. Bacterial cells lack a membrane bound nucleus. Their genetic material is naked within the cytoplasm. A wall located outside the cell membrane provides the cell support, and protection against mechanical stress or damage from osmotic rupture and lysis. The major component of the bacterial cell wall is peptidoglycan or murein. This rigid structure of peptidoglycan, specific only to prokaryotes, gives the cell shape and surrounds the cytoplasmic membrane.

Peptidoglycan

Peptidoglycan is a huge polymer of disaccharides (glycan) cross-linked by short chains of identical amino acids (peptides) monomers. The peptidoglycan layer in the bacterial cell wall is a crystal lattice structure formed from linear chains of two alternating amino sugars, namely N-acetylglucosamine (GlcNAc or NAG) and N-acetylmuramic acid (MurNAc or NAM). The alternating sugars are connected by a β-(1,4)-glycosidic bond. They are transported across the cytoplasmic membrane by a carrier molecule called bactophenol. Each MurNAc is attached to a short (4- to 5-residue) amino acid chain, containing L-alanine, D-glutamic acid, *meso*-diaminopimelic acid, and D-alanine in the case of a Gram-negative bacterium or L-alanine, D-glutamine, L-lysine, and D-alanine with a 5-glycine interbridge between tetrapeptides in the case of a gram-positive bacterium. Peptidoglycan is one of the most important sources of D-amino acids in nature.Cross-linking between amino acids in different linear amino sugar chains occurs with the help of the enzyme transpeptidase. The specific amino acid sequence and molecular structure vary with the bacterial species.

From the peptidoglycan inwards all bacterial cells are very similar. The cell wall provides important ligands for adherence and receptor sites for viruses or antibiotics. The peptidoglycan layer is substantially thicker in Gram-positive bacteria (20 to 80 nanometers) than in Gram-negative bacteria (7 to 8 nanometers).

Transpeptidation

Transpeptidation (the process of cross-linkage of peptide chains to produce the insoluble, strong mesh of peptidoglycan) process is catalyzed by a class of transpeptidases known as penicillin binding proteins (PBPs) that links D-Ala$_{(4)}$ from one stem peptide to the free amino group of AA$_{(3)}$ on another stem peptide and results in a 3-dimensional structure that is strong and rigid. A critical part of the process is the recognition of the D-Ala-D-Ala sequence of the NAM peptide side chain by the PBP. During transpeptidation, the D-Ala-D-Ala bond of one stem peptide is first cleaved and an enzyme-substrate intermediate

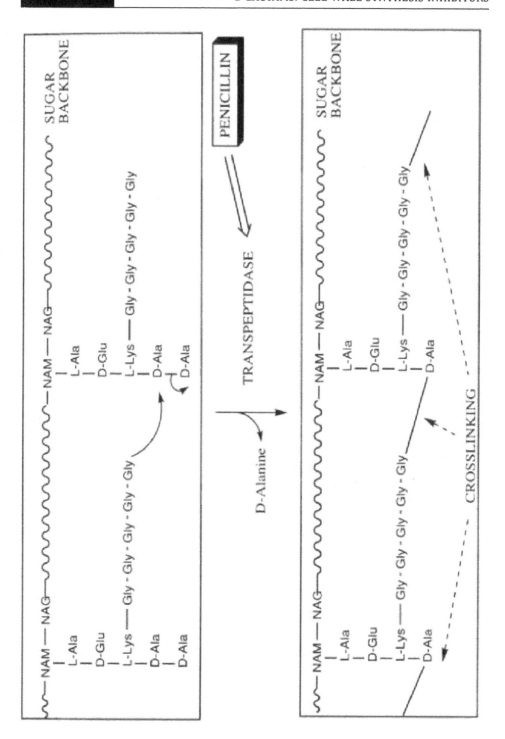

is formed, with the concomitant release of the terminal D-Ala. The cleavage reaction provides the energy necessary for the transpeptidation reaction, which occurs outside the cytoplasmic membrane in the absence of energy donors such as ATP. A second step involves the transfer of the peptidyl moiety to an acceptor. This acceptor is the non-alpha amino group of the dibasic amino acid in a second stem peptide. The reaction results in the formation of a new peptide bond between the penultimate D-alanine of a donor peptide and an amino group of the cross bridge of an acceptor peptide.

Factors influencing activity of B-lactam antibiotics

a) Beta-lactam antibiotics are most active during the logarithmic phase of bacterial growth because they act against rapidly growing and replicating cells. They have little effect on formed bacterial cell wall.

b) The efficacy of β-lactam antibiotics is time dependent. Therefore, for maximal efficacy, the plasma tissue concentrations must be maintained above the MIC for long periods of time.

c) The efficacy of β-lactam antibiotics is more in an isotonic environment (isotonic to the host and hypotonic to the organism) because it causes rapid movement of fluid into the hypotonic bacterial cell that results in rapid osmotic lysis. The efficacy of penicillins generally decreases in a hypertonic environment due to deficient osmotic lysis.

d) The efficacy of B-lactam antibiotics may decrease in chronic infection or when density of bacterial population is more. This probably occurs due to slow growth of microorganisms in a chronic infection or presence of greater number of relatively resistant organisms in a large population.

Antimicrobial spectrum

Penicillins differ in their antibacterial spectra. In general, gram-positive micro-organisms are more susceptible to penicillins because their cell wall is located close to the cell surface and is readily transversed by penicillins. In contrast, gram-negative bacteria contain a capsule and outer lipopolysaccharide membrane surrounding the cell wall that prevents many penicilllins to reach cell wall and target PBPs. Some small hydrophilic broad spectrum penicillins, however, diffuse through the porin channels to reach the outer membrance where third stage of peptidoglycan synthesis takes place. Accordingly, narrow spectrum penicillins (e.g., penicillin G) are active against only Gram-positive organisms and broad-spectrum penicillins are active against large number of Gram-positive (mainly) and Gram-negative organisms. Many bacteria are inherently insensitive to penicillins because in them the target PBPs are located deeper under the lipoprotein barrier where β-lactam antibiotics are unable to penetrate. Additionally, some micro-organisms may be intrinsically resistant to penicillin because of structural difference in the PBPs that are target for these drugs.

Bacterial Resistance

Acquired resistance to penicillins is a serious problem and occurs mainly through transfer of plasmids (also chromosomes). It may be acquired through following mechanisms.

1. **Beta-lactamase activity:** Inactivation of β-lactam antibiotics by elaboration of β-lactamase (penicillinase) enzymes by bacteria is the most important and clinically significant mechanism of acquiring resistance. The β-lactamases are a family of enzymes produced by many Gram-positive and Gram-negative bacteria that hydrolyse the cyclic bond of β-lactam ring resulting in loss of bactericidal activity. Several types (at least 6) of β-lactamase are produced by bacteria, some are active exclusively against penicillins (called penicillinases), some are active against cephalosporins (called cephalosporinases), and others are active equally against both penicillins and cephalosporins (broad-spectrum β-lactamases).

2. **Decreased permeability to drug:** Decreased penetration of the β-lactam antibiotics through outer membrane prevents the drugs from reaching the target penicillin-binding proteins. Some Gram-negative bacteria become resistant to penicillins by loss or alteration of aqueous channels (porins) in the outer membrane. Active efflux of antibiotics through cell membranes also serves as another mechanism of acquiring resistance.

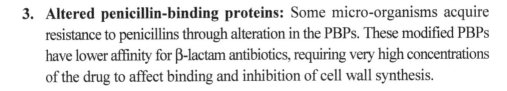

3. **Altered penicillin-binding proteins:** Some micro-organisms acquire resistance to penicillins through alteration in the PBPs. These modified PBPs have lower affinity for β-lactam antibiotics, requiring very high concentrations of the drug to affect binding and inhibition of cell wall synthesis.

Pharmacokinetics

The pharmacokinetics of many penicillins differ substantially. The route of administration is determined by the stability of drug in gastric acid and by the severity of infection. Although the pharmacokinetic properties of individual drugs are discussed separately, following generalisation can be useful.

The penicillins are weak acids, which favour oral absorption. However, many penicillins destroyed by the gastric acid and thus, cannot be given orally. Absorption occurs from the upper gastro intestinal tract. But degree and rate of absorption differs greatly among various penicillins. After absorption, penicillins are widely distributed throughout the body. Penicillins are bound to plasma proteins to variable extent (20-80%). Penicillins do not undergo significant biotransformation and are generally excreted unchanged.

Drug interactions

Combinations of penicillins with aminoglycosides can have synergistic effect, whereas use of penicillins with bacteriostatic drugs like Chloramphenicol, Tetracyclines and Erythromycin may produce antagonistic effects and should not be mixed *in vitro*.

I. Narrow–Spectrum Penicillins

Narrow-Spectrum Beta-Lactamase Sensitive Penicillins

Narrow-spectrum β-lactamase sensitive penicillins include naturally occurring penicillin G in its various pharmaceutical forms and a few semi-synthetic/ biosynthetic acid stable penicillins such as penicillin V and phenethicillin. Semi-synthetic penicillins are produced by incorporation of specific precursors in mould cultures (e.g., penicillin V) or made by modification of penicillin G or 6-

amino penicillanic acid (6-APA)(e.g., phenethicillin). Penicillins of this group are active against mainly Gram- positive bacteria. All these penicillins are susceptible to β-lactamase inactivation.

Acid Susceptible Penicillins/Natural Penicillins

Penicillin G/Benzylpenicillin

Penicillin G (Benzylpenicillin) is a natural penicillin obtained from fermentation of the mould *Penicillium chrysogenum*. It was the first of the penicillins clinically used and still remains an important and useful antibacterial agent. Penicillin G is inactivated by gastric acid so is not usually administered by mouth. It is also susceptible to inactivation by B-lactamases. The potency of penicillin G is usually expressed in terms of Units. The International Unit (I.U) of penicillin is the amount of activity present in 0.6 µg of the crystalline sodium salt of penicillin G.

Penicillin V/Phenoxymethyl Penicillin

Penicillin V (Phenoxymethyl penicillin) is a variant of penicillin G obtained by adding 2-phenoxyethanol to the Penicillium culture using yeast autolyzate as a source of nitrogen. The phenoxymethyl group on penicillin V imparts more acid stability on oral administration. It has a similar antibacterial spectrum to penicillin G, but is less active. After oral administration, penicillin V is better absorbed than the penicilliln G, but absorption is unpredictable and still relatively low for most systemic infections. It is mostly used as potassium salt (penicillin V potassium).

Narrow-Spectrum Beta-Lactamase Resistant Penicillins

This group of penicillins include several semi-synthetic antibiotics produced by chemical modification of the penicillin nucleus. The chemical modification hinders the access of β-lactamase enzymes produced by resistant gram-positive organisms, to the β-lactam ring making the drugs resistant to inactivation by this group of enzymes. Eg. Cloxacilin, Oxacillin, Dicloxacilin.

II. Broad-spectrum Penicillins

Broad-spectrum penicillins include some semi-synthetic penicillins and their precursors that are active against many Gram-positive and Gram-negative bacteria. They are all destroyed by β-lactamases produced by various bacteria. Many members of the group are acid stable and are administered either orally or parenterally. Eg. Ampicillin, Amoxycillin.

iii. Extended-Spectrum Penicillins/Anti-Pseudomonal Penicillins

Extended-spectrum penicillins include several semi-synthetic broad-spectrum penicillins that are effective against several bacteria including *Pseudomonas aeruginosa* that is resistant to ampicillin and its congener amoxycillin, Eg. Carbenicillin

iv. Potentiated Penicillins/Beta Lactamase Protected Penicillins

Potentiated penicillins are combinations of a β-lactamase inhibitor and a broad spectrum β-lactamase sensitive penicillin. Such combinations produce significant potentiative effect because the active penicillin is protected from enzymatic hydrolysis. Eg. Amoxycillin-Clavulanic Acid

Cephalosporins

Cephalosporins are derived semi-synthetically from 'Cephalosporin-C' obtained from a fungus, *Cephalosporium*. Initially 3 natural cephalosporins *viz.,* cephalosporins N, and C were isolated from a culture of *Cephalosporium acremonium* fungus by Brotzu in 1948.

Cephalosporins are bactericidal (kill bacteria) and work in a similar way to penicillins. They bind to and block the activity of enzymes responsible for making peptidoglycan, an important component of the bacterial cell wall. They are called broad-spectrum antibiotics because they are effective against a wide range of bacteria.

Since the first cephalosporin was discovered in 1945, scientists have been improving the structure of cephalosporins to make them more effective against a wider range of bacteria. Each time the structure changes, a new "generation" of cephalosporins are made.

Classification of Cephalosporins

Cephalosporin drugs are divided into different generations depending upon their microbial spectrum.

1. First generation

2. Second generation

3. Third generation

4. Fourth generation

First-generation cephalosporins are predominantly active against Gram-positive bacteria, and successive generations have increased activity against Gram-negative bacteria.

First Generation

The optimum activity of all first generation cephalosporin drugs is against Gram-positive bacteria such as Staphylococci and Streptococci. They also have little Gram-negative spectrum. E.g. Cefazolin, Cephalothin, Cephapirin, Cephalexin, Cefadroxil and Cephradine.

Second Generation

The drugs that come under second generation have more spectra against Gram-negative bacteria (*Haemophilus influenzae, Enterobacter aerogenes*) in comparison to the first generation. Their Gram-positive spectrum is less than the first generation. E.g. Cefamandole, Cefuroxime, Cefoxitin, Cefotetan, Cefmetazole, Cefaclor, Cefprozil, Cefpodoxime and Loracarbef

Third Generation

Third generation Cephalosporin drugs are broad spectrum and are effective against both Gram-positive and Gram-negative bacteria. However, their optimum activity is against Gram-negative bacteria. E.g. Cefotaxime, Ceftriaxone, Ceftizoxime, Ceftazidime, Cefoperazone and Cefixime

Fourth Generation

These are extended spectrum antibiotics. They are resistant to beta lactamases. E.g. Cefipime

Carbapenem

Carbapenems are a class of β-lactam antibiotics with a broad spectrum of antibacterial activity. Carbapenems inhibit bacterial cell wall synthesis by binding to the penicillin binding proteins and interfering with cell wall formation. They are extremely resistant to β-lactamase enzymes, making them very useful in treating bacterial infections where β-lactamase is produced that makes other β-lactam antibiotics ineffective. Carbapenem antibiotics were originally developed from thienamycin, a naturally-derived product of *Streptomyces cattleya.*

Monobactam

Monobactams are β-lactam compounds in which the β-lactam ring is alone and not fused to another ring as opposed to other β-lactams, which have at least two rings. Monobactams only have activity against Gram-negative bacteria. Aztreonam is an economical, broad-spectrum antibiotic for plant tissue culture used especially for controlling *Pseudomonas aeruginosa* pathogens. The only commercially available monobactam antibiotic is aztreonam. Aztreonam, the first marketed monobactam, has activity against most aerobic Gram-negative bacilli including *P. aeruginosa.* They have no cross-hypersensitivity reactions with penicillin but, like penicillins they can trigger seizures in patients with history of seizures.

Uses

Belonging to the group of β-lactam antibiotics, ampicillin is able to penetrate Gram-positive and some Gram-negative bacteria. Even though this is a commonly used antibiotic family in humans and other warm blooded animals, this has shown to be generally ineffective in fish (especially Penicillin). The only effectiveness comes in the treatment of some fungal-like eye infections.

β-Lactamase inhibitors

β-lactamase inhibitors are agents that bind to and inactivate b-lactamase enzyme produced by many Gram-positive and Gram-negative bacteria. These drugs are formulated with the penicillins to protect them from enzymatic degradation.

Clavulanic Acid

Clavulanic acid is β-lactam ring containing compound obtained from *Streptomyces clavuligens*. It inhibits a variety of β-lactamases produced by bacteria.

Mechanism of action

Clavulanic acid contains a β-lactam ring in its structure. But, unlike β-lactam antibiotics it does not produce bactericidal action. Instead, having a β-lactam ring and structural similarities with penicillins, it binds to and inactivates lactamase enzyme produced by certain bacteria. During the inactivation process, Clavulanic acid initially binds to the active site of β-lactamase enzyme to form an acyl-enzyme complex. The complex then follows one of the two reaction pathways, both of which result in, inactivation of the enzyme. In the first reaction pathway (pathway 1), the acyl-enzyme complex undergoes a rearrangement process that results in a transiently inhibited enzyme. In the second reaction pathway (pathway 2), first there is an attack of an active site amino group on the acyl-enzyme complex followed by a second modification (rearrangement) process that results in an irreversibly inactivated enzyme. The inactivation of enzyme occurs gradually

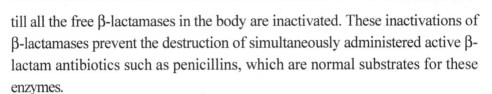

till all the free β-lactamases in the body are inactivated. These inactivations of β-lactamases prevent the destruction of simultaneously administered active β-lactam antibiotics such as penicillins, which are normal substrates for these enzymes.

Uses

The drugs have a poor activity against *Vibrio* sp. and *Yersinia ruckeri*. It is used to treat furunculosis in salmonids, Pasteurellosis in sea bass, sea bream, Edwardsiellosis in catfish and eels.

8

SULPHONAMIDES

Sulphonamides (Sulfonamides) are a group of synthetic organic chemicals with chemotherapeutic activity. They were the first chemotherapeutic agents used systemically for prevention and treatment of various bacterial infections. Although a large number of sulphonamides are currently available in market, their clinical use has declined mainly due to emergence of bacterial resistance and availability of better antibacterial agents.

History

The antibacterial activity of sulphonamides was first discovered by Domagk in 1935, who demonstrated that prontosil, an azo dye, is effective against streptococcal infection in mice. Noble Prize in medicine was awarded for this discovery in 1938. Soon thereafter it was discovered that the antibacterial activity of prontosil is due to the release of an active metabolite, the sulphanilamide. Thereafter, sulphanilamide was synthesised and introduced in market as the first sulphonamide for antibacterial use. Since then, a large number of sulphonamides with variable pharmacokinetics and antibacterial spectra have been developed for use in human and veterinary medicines.

Chemistry and structure-activity relationship

The term sulphonamide (sulpha drug) is commonly used as a generic name for all derivatives of sulphonamides that possess a common chemical nucleus, which is closely related to para-aminobenzoic acid (PABA), an essential member of vitamin B complex. This nucleus is essential for antibacterial activity. The sulphonamide nucleus possesses two amine groups, which has been designed as N^4, which mainly governs solubility, potency and pharmacokinetic property. Broadly, following generalisations can be made:

a) The presence of para-amino group (N^4) is essential for the antibacterial activity, but it can be substituted by groups that *in vivo* are converted into amine groups.

b) Substitution at amide group (N^1) can have variable effects. Substitution at N^1 by heterocyclic aromatic nuclei yields highly potent compounds such as sulphamerazine, sulphadiazine, and sulphadimidine.

c) The sulphamoyl group ($-SO_2NH_2$) is not essential as such; only that sulphur should be directly linked to the benzene ring.

d) Any substitution on the benzene ring results in loss of antibacterial activity.

e) Acetylation at N^1 may not alter the chemotherapeutic activity (or may decrease), but such compounds become water soluble and are less toxic. These compounds may be used for renal or eye infections e.g., sulphacetamide.

f) With the exception of some sulphonamides (pyrimidine sulphonamides like sulpha-merazine and sulphadiazine), acetylation at N^4 decreases the water solubility and enhances chances of renal toxicity. The antibacterial activity may be abolished by N^4 acetylation.

Chemical properties and solubility

Sulphonamides are weak organic compounds. Most of these are relatively insoluble in water, but their sodium salts are water-soluble. Some sulphonamide solutions have pH values more than 9(9-10), which prohibit their extra vascular use due to strong irritant properties to tissues. In acid urine, sulphonamides may form crystals because of their decreased solubility. The N^4 acetylated sulphonamides derivatives are relatively insoluble in acidic urine and some may even precipitate in the renal tubules of species having acid urine (e.g., carnivores) leading to crystalluria and renal failure. Solubility of sulphonamides is affected (increased) by presence of another sulphonamide in the solution because they follow law of independent solubility (i.e., in a mixture of sulphonamides, each sulphonamide exhibits its own solubility in solution). Therefore, combination of 2 or more sulphonamides is occasionally used to increase solubility and efficacy (additive effect) and to decrease toxicity (less crystallisation in urine).

CLASSIFICATION

Sulphonamides are usually classified into following groups depending on their type and duration of action:

I. Systemically acting sulphonamides

1. Short acting sulphonamides (duration<12 hours)

    e.g., sulphadiazine, sulphisoxazole, sulphamerazine, sulphachlorpyri-dazine,sulphathiazole, sulphanilamide, sulphapyridazine and sulphasomidine.

2. Intermediate acting sulphonamide (duration 12-24 hours)

    e.g., sulphadimidine, sulphamethoxazole, sulphamoxide and sulphaphe-nazole.

3. Long –acting sulphonamides (duration 24-48 hours)

    e.g., sulphadimethoxine, sulphae-thoxypyridazine, sulphamethoxy-pyridazine and Sulphabromone-thazine.

4. Ultra-long acting sulphonamides (duration>48 hours)

    e.g., sulphadoxine and sulphame-thopyrazine.

II. Locally acting sulphonamides

1. Gut-acting sulphonamides

    e.g., succinysulphathiazole, phthalylsulphathiazole, phthalylsul-phacetamide, sulphaguinidine and sulphasalazine.

2. Topically acting sulphonamides

    e.g., Sulphacetamide, Mafenide and silver sulphadiazine.

Mechanism of action

Sulphonamides are structural analogues of para-aminobenzonic acid (PABA). PABA is an essential component of folic acid (pteroylglutamic acid). Mammals require preformed folic acid in their diet, but many bacteria utilize PABA for the synthesis of folic acid, which in turn is utilized for the synthesis of some essential macromolecules like thymidine, purines, methionine and glycine that are vital for bacterial multiplication and growth. When a sulphonamide is given, due to its structural similarity with PABA, it blocks utilization of PABA with pteridine residue to form dihydropteroic acid. Inhibition of dihydropteroic acid further inhibits synthesis of dihydrofolic acid (dihydrofolate) and then tetrahydrofolic acid (tetrahydrofolate), the reduced form of folic acid. Also being chemically similar to PABA, the sulphonamide may itself get incorporated to form a defective folate that is metabolically harmful to the bacteria. As a result, bacteria cease to multiply.

The sulphonamides are primarily bacteriostatic and have no effect on resting bacteria. There is a lag phase, time differences between action of sulphonamide and inhibition of bacterial multiplication, during which bacteria utilize existing folic acid synthesised prior to exposure to sulphonamide. Sulphonamides are more effective in acute stage of infection because in acute infections, bacteria multiply at a much faster rate utilizing large amounts of PABA. Also in acute stage, host's immune system is very active and rate of phagocytosis is sufficient to remove the affected bacteria. Since PABA is utilized for the synthesis of thymidine, purines and methionine, the presence of these substances in environment may decrease efficacy of sulphonamides. Pus and tissue debris may also contain large amount of these substances as break down products, therefore, sulphonamides are less effective in presence of necrotic tissues, pus and tissue extracts. Pus is also rich in PABA. Sulphonamides generally produce selective toxicity on target microorganisms and do not affect animal cells because mammalian cells utilize the preformed folic acid and do not synthesise it in the body as in the bacteria. Accordingly, sulphonamides also have no/little effect on those microbial organisms, which utilizes the preformed folic acid.

Sulphonamide antagonism

The antibacterial action of sulphonamides is antagonised by the supply of metabolities whose synthesis is inhibited by them. These include:

a) Presence of PABA or drugs whose metabolism yields PABA

b) Supply of vitamin B complex such as niacin, folic acid, choline and amino acids like glutamic acid and methionine.

c) Some proteins such as gelatin, albumin, and peptone that bind with sulphonamides and reduce their availability.

d) Products of cell and tissue death, especially pus. They supply products that neutralize sulphonamides or act as non-vascular barriers and reduce diffusion of drugs.

Synergists of sulphonamides

Sulphonamides show synergistic action with some diaminopyrimidines such as trimethoprim. Trimethoprim is a potent and selective competitive inhibitor of dihydrofolate reductase enzyme in susceptible micro organisms, the enzyme required to reduce dihydrofolate to tetrahydrofolate. Thus combination of sulphonamide and trimethoprim produces sequential blocks in the synthesis of tetrahydrofolate, the reduced form of folic acid that is required for one carbon transfer reactions.

Antimicrobial spectrum

Sulphonamides are broad-spectrum antimicrobial drugs. They are primarily bacteriostatic with moderate to good activity against many Gram-positive and Gram-negative bacteria including *Actinomyces* spp., *Bacillus* spp., *Pasteurella* spp., Enterobacteriaceae and some *Clostridium* spp. Some bacteria like Pseudomonas are resistant to sulphonamides. Also resistant to sulphonamides are Mycobacterium and Mycoplasma. Some protozoa like Coccidia and Toxoplasma are sensitive to sulphonamides. Action of sulphonamides on many above-mentioned organisms is more pronounced in urinary tract infections because in urine, sulphonamides may attain bactericidal concentration.

Bacterial resistance

Bacterial resistance is fast developing to sulphonamides and a large number of microorganisms found initially sensitive to sulphonamides are now insensitive to them. Acquired bacterial resistance to sulphonamides is believed to originate by random mutation and selection or by transfer of resistance by plasmids. The bacterial resistance may develop due to one or more of the following causes:

a) An alteration in the bacterial enzyme that utilise PABA, the dihydroate synthase enzyme may have low affinity for sulphonamides.

b) An increased capacity of bacteria to destroy or inactivate sulphonamides.

c) An increased production of PABA or essential metabolities by bacteria.

d) Adoption of alternate pathway for synthesis of essential macromolecules by bacteria.

e) Decreased drug permeability into the bacterial cell or active efflux of drug from the target bacteria.

The bacterial resistance to sulphonamides, particularly when developed *in vivo*, is usually persistent and irreversible. Although the acquired resistance to sulphonamides generally does not involve cross-resistance to other antimicrobial drugs, bacterial resistance to one sulphonamide generally provides resistance to all sulphonamides.

Steps to reduce bacterial resistance

Following points may help reduce emergence of resistant strains of bacteria to sulphonamides.

a) Sulphonamide therapy should be initiated in acute stage of disease.

b) Sulphonamide therapy should be continued till complete recovery from infection occurs.

c) Initial doses of sulphonamide may be kept high to establish therapeutic blood concentration. This may be followed by smaller maintenance doses.

d) Sulphonamides should be used only when the infection is from sulphonamide sensitive microorganisms. Indiscriminate use of sulphonamides should be avoided.

Pharmacokinetics

Except for sulphonamides especially designed for local action (gut-acting and topically acting), this class of antimicrobial drugs have a common pharmacokinetic pattern:

Renal toxicity: Renal toxicity occurs due to precipitation of the sulphonamide in the glomerular filtrate leading to crystallization. Chances of crystalluria are more in dehydrated animals as the urine may contain drug concentrations above solubility limits resulting in crystal formation. Clinical use of less water soluble sulphonamides or rapidly excreting sulphonamides (e.g., sulphathiazole) have greater tendency to get precipitated in the renal tubules.

Hypersensitivity reactions: Allergic skin rashes are frequent complication of sulpha therapy. Cutaneous eruption may occur and are more common with long acting sulphonamides, so they have not become popular.

Others effects: Several other acute effects may occur in sulphonamide treated animals. These include anorexia and abdominal discomfort. Chronic toxicity occurs in dose-dependent manner and may be seen after prolonged administration of sulphonamides. Some of the effects, however, occur after a short-term exposure.

Hypoprothrombinaemia: Prolonged administration of certain sulphonamides may lead to vitamin k deficiency due to inhibition of enzyme vitamin K epoxide reductase. This results in hypoprothrombinaemia and deficiency of other vitamin K dependent clotting factors causing prolongation of bleeding and clotting time. It is mostly seen in poultry with sulphaquinoxaline.

Inhibition of carbonic anhydrous enzymes: Prolonged administration of sulphonamides inhibit carbonic anhydrous enzyme resulting in accumulation of carbon dioxide leading to acidosis.

Clinical uses

- It is the drug of choice particularly in infections of the internal organs caused by the organisms i.e. Actinomyetes, cocci, and many Gram-positive and a

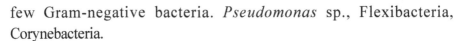

few Gram-negative bacteria. *Pseudomonas* sp., Flexibacteria, Corynebacteria.

- It can be used for fin and tail rot, mouth fungus, Columnaris, and hemorrhagic septicaemia.

- Since Sulfas are more effective at higher pH levels, this can be a good choice for livebearer or goldfish fin rot (as well as any other fish generally kept at a higher pH)

- Sulphonamides are administered by various routes depending on the type of salt, solubility, action required, and species. Since sulphonamides pass well from the gut to the blood, mixing it into the feed is its most effective application. The poor solubility of the drug makes its use difficult in a long bath.

SYSTEMICALLY ACTING SULPHONAMIDES

1. Short-acting Sulphonamides

Short acting sulphonamides are rapidly absorbed and rapidly eliminated and have short duration of action (4-8 hours). Some of these drugs(e.g., sulphisoxazole and sulphasomidine) are specifically used for urinary tract infection because they are rapidly eliminated in urine, are more soluble in urine pH, and undergo less acetylation metabolism.

Sulphadiazine

Sulphadizine (Sulphapyrimidine) is the prototype of the short-acting sulphonamides. It exists as white or slightly yellow powder that is sparingly soluble in water, alcohol, and acetone, but freely soluble in solutions of potassium and sodium hydroxides and in ammonia in water. It is rapidly absorbed from GI tract and excreted readily by kidney in both free and acetylated (15-40%) forms. It is less protein bound (14%), which facilities diffusion into tissues. Sulphadiazine was widely used in fishes. Presently, it is mostly used in the potentiated form with trimethoprim and may be combined with other sulphonamides such as sulphamerazine and sulphadimidine. The acetylated derivative of sulphadiazine is less soluble in urine, so crystalluria is possible.

Sulphisoxazole

Sulphisoxazole (Sulphafurazole) is a water-soluble short-acting sulphonamide. Its solubility at pH 5 to 7 is so high that neither the compound itself nor its N4-acetyl derivative is deposited in kidneys. Therefore, chances of crystalluria are unlikely, even in acid urine. Sulphisoxazole is rapidly absorbed and rapidly excreted in urine. Its concentration in the urine greatly exceeds that in blood, therefore it is mostly used for urinary tract infection, particularly in small animals. It is also suitable for some systemic infections because of lower propensity to cause crystalluria.

Sulphamerazine

Sulphamerazine (Sulphamethyldiazine) is a short acting systemic sulphonamide. It is soluble in water and readily soluble in solutions of potassium, sodium, or ammonium hydroxides. Sulphamerazine is rapidly absorbed after oral administration and rapidly excreted in the urine. It is used both orally and parenterally to treat a variety of infections. Sulphamerazine may be used alone or in combination with other sulphonamides (e.g.,sulphadimidine, sulphadiazine).

It was one of the earliest sulphonamides to be the subject of a market authorization for fish. Sulphamerazine has been approved in the USA for use in farmed fish with a dose regimen of 200 mg/kg/day for 14 days.

Sulphachlorpyridazine

Sulphachlorpyridazine is a short to intermediate acting systemic sulphonamide. It is highly soluble in urine at normal pH. It is readily absorbed from GI tract and is readily excreted in urine. It is indicated mainly for the enteric treatment caused or complicated by *E.coli*. It is also used parenterally as a general purpose sulphonamide.

Sulphathiazole

Sulphathiazole is a short-acting sulphonamide that is occasionally used by oral route and is also included in some parenteral formulations. It is less soluble in water and more toxic than other short-acting sulphonamides. It is safe when used as the phthalyl derivative (phthalylsuphathiazole). In veterinary practice, sulphathiazole is mainly formulated with chlortetracycline and procaine penicillin G.

2. Intermediate-acting Sulphonamides

Intermediate-acting sulphonamides include drugs that are rapidly absorbed but relatively slowly excreted from the body.

Sulphadimidine

Sulphadimidine (Sulphamethazine, sulphamezathine, sulphadimerazine, sulphamidine) is a widely used intermediate to long-acting sulphonamide. It is rapidly absorbed after oral administration but is relatively slowly excreted form the body so that therapeutic levels may be maintained for up to 24 hours with a single dose. About 70% of the sulphadimidine is protein bound. The metaboliltes of sulphadimidine are highly soluble, so the chances of crystal formation in renal tubules are less. Clearance of sulphadimidine and its metabolites is dose-dependent.

Sulphamethoxazole

Sulphamethoxazole (Sulphamethoxizole, sulphamethylisoxazole, sulphisomezole) is chemically related to sulphisoxazole, but it shows intermediate duration of action with relatively slower oral absorption and urinary excretion. It is administered orally for both urinary and systemic infections. It produces high percentage of an acetylated metabolite, which is relatively insoluble and may cause crystalluria.

Sulphamoxole

Sulphamoxole (sulphadimethyloxazole) is a close congener of sulphisoxazole and sulphamethoxazole. It is soluble in water and unlike many sulphonamides its solubility decreases with increasing pH value. Its antibacterial properties are similar to sulphamethoxazole and is employed when a sulphonamide alone has to be used for urinary or respiratory tract infection.

3. Long-Acting Sulphonamides

Long-acting sulphonamides are well absorbed, but are slowly excreted from the body. They are highly bound to plasma proteins and have high renal tubular reabsorption. Their half-lives may range from 24 to 48 hours. Due to the tissue accumulation, they are also more toxic.

Sulphadimethoxine

Sulphadimethoxine is a long-acting sulphonamide. It is less soluble in water and slightly soluble in alcohol. After oral administration, it is readily absorbed from GI tract. Sulphadimethoxine is well distributed in body. The concentrations being bile > intestine > liver > blood > skin > kidney > spleen > gills > muscle. Little drug excretes through the gills or urine and high concentration in the bile result in recirculation, so slow elimination is expected. The N^4 acetyl derivatives of sulphadimethoxine undergoes significant reabsorption that is primarily responsible for its long elimination half-life.

Sulphaethoxypyridazine

Sulphaethoxypyridazine is a long-acting sulphonamide with systemic antibacterial activity. It is readily absorbed after oral administration and is extensively bound to plasma protein. It is used mainly for the treatment of large animal infections.

4. Ultra-long acting Sulphonamides

Some sulphonamides have very long–duration of action and are repeated once or twice in a week. These include sulphadoxine, sulphamethoxypyridazine, and sulphamethopyrazine.

Sulphadoxine

Sulphadoxine is an ultra-long acting sulphonamide. Its action usually lasts for about a week after a single dosing because of its high plasma protein binding and slow urinary excretion. After oral administration, it attains low plasma concentration mainly due to extensive protein binding. Therefore, it is not considered useful for the treatment of acute bacterial infection.

Sulphamethopyrazine

Sulphamethopyrazine is an ultra-long acting sulphonamide with properties similar to those of sulphadoxine. It is mostly marketed in fixed – dose combination with pyrimethamine as an antimalarial drug. Its aquaculture use is limited.

II. Locally-acting Sulphonamides

1. Gut-acting sulphonamides

Gut-acting sulphonamides are very poorly absorbed from the GI tract and are excreted largely unchanged in faeces. They provide high concentration of the drug in the GI tract due to low absorption and may also release active sulphonamide due to bacterial hydrolysis.

Succinylsulphathiazole and Phthalyl-sulphathiazole

Succinylsulphathiazole and Phthalyl-sulphathiazole are a group of sulphonamides that are poorly absorbed following oral administration. Both drugs lack antibacterial activity *in vivo* because they are primarily prodrug forms of sulphonthiazole. On oral administration, Succinylsulphathiazole and Phthalyl-sulphathiazole in the GI tract release Succinic or pthalic acid and sulphithiazole, the latter being the effective drug.

Phthalylsulphacetmide

Phthalylsulphacetmide is a gut acting sulphonamide that releases active sulphacetamide in the large intestine.

Sulphaguinidine

Sulphaguinidine is another gut acting sulphonamide. It is absorbed poorly when administered orally because it is consistently ionized at the pH of GI contents.

2. Topically acting sulphonamide Sulphacetamide

Sulphacetamide is probably the only sulphanamide whose sodium salt has neutral pH and is non-irritant to tissues therefore, it is used as a topical treatment .

Silver sulphadiazine

It is used topically to inhibit pathogenic bacteria and fungi including those resistant to other sulphonamides (Pseudomonas). On topical application, it slowly releases silver ions, which appear to be largely responsible for antimicrobial action.

Sulphonamides used for treating fish

Antibiotic	Route of administration	Dosage	Withdrawl time	Adverse effects
Sulphamerazine Sulphadimidine Sulphadimethoxine Sulphisoxazole	Oral	200 mg/kg	3-4 weeks	Sterility and nephrotoxicity after prolonged use

Triple sulfa (Sulfamerazine, Sulfamethazine, Sulfathiazole)

A relatively broad spectrum antibacterial medication suitable for fin and tail rot, mouth fungus, Columnaris (mild to moderate infections only) and hemorrhagic septicemia (although not effective to Aeromonas infections of the gut).

Triple Sulfa is often a good choice when there is an infection caused by a scrape, abrasion, or similar. While not often the first choice for fish eye issues, including pop-eye, it can be a first choice or a second choice if the first choice fails in treating such maladies involving the eye. This would not only include and in-tank treatment but use as part of a bath treatment too. It can be used in combination with Malachite Green (especially effective in combination with MG at half strength when treating Ich in scaleless fish) or Acriflavin (do not combine with copper sulfate).

Diaminopyrimidines (Potentiated Sulphonamides)

Potentiated sulphonamides are the combinations of sulphonamides with 2,4-diamopyrimidine and related agents. The combinations commonly used are that of sulphamethoxazole and trimethoprim (co-trimaxazole), sulphadiazine and trimethoprim (co-rimazine), sulphamethoxine and ormetoprim.

Mechanism of action

The combination of sulphonamide with 2, 4-diaminopyrimidine derivative results in synergistic effect *via* blockade of sequential stages in the synthesis of tetrahydrofolate. Sulphonamide competitively antagonises PABA and inhibits its incorporation into dihydropteroic acid and 2,4-diaminopyrimidine drug prevents reduction of dihydrofolic acid to tetrahydrofolic acid by competitively inhibiting the enzyme dihydro-folate reductase. Thus, selective toxicity for microorganisms is achieved in two ways and the combination becomes bactericidal (sulphonamides and 2, 4-diaminopyrimidine when used separately are bacteriostatic).

Apart from the bactericidal effect, there are certain advantages of using potentiated sulphonamides. The combination of sulphonamide and 2,4-diaminopyrimidine reduces by several fold the MIC of both drugs against a wide variety of pathogenic organisms. This reduces the dose of sulphonamide required for therapeutic effect and also chances of dose-dependent adverse effects. The combination of sulphonamide with 2, 4-diaminopyrimidine also broadens the antibacterial spectrum and reduces chances of bacterial resistance of drugs. Furthermore, 2, 4-diaminopyrimidine selectively inhibits the dihydrofolate reductase enzyme of microorganisms with no/negligible effect on the enzyme of mammals. This is vitally important because mammals also utilize dihydrofolate reductase enzyme in folic acid metabolic pathways.

Clinical Uses

Combining trimethoprim with a sulfonamide offers a significant improvement in efficacy. Trimethoprim is a bacteriostatic antibiotic effective for many aerobic Gram-negative bacterium including *Pseudomonas* and *Aeromonas*. Since *Pseudomonas* and *Aeromonas* are common causes of opportunistic fin rot in fish, this drug or a combination that may be a good alternative treatment. In fact, when combined with some Sulfa based medications, it produced a synergism or addition in 85% (similar to how Kanamycin and Nitrofurazone produce a synergism that treats *Columnaris* and *Aeromonas* than when treated alone, often results in failure). However Trimethoprim has no proven effectiveness for anaerobic infections, so if the causes of an *Aeromonas* is anaerobic (which most are), then this or antibiotics containing Trimethoprim would be a poor choice. Especially useful for fin and tail rot. Since pathogens develop resistance very rapidly, the medication should be used only once every six months.

Commercially available preparations include

Drylin, Eusaprim, Borgal solution (7.5%), and Cotrimstad-forte. Drylin and Eusaprim are older preparations with low concentrations of active ingredient. Today the most commonly used preparations are Bogai and Cotrimstada-Forte.

Contraindications

- Can be very harsh to nitrifying bacteria in an aquarium, do not over dose and use only in well established aquariums.

- Can cause Thrombocytopenia (lowering of blood platelets), so this is a poor choice if fish have large wounds or are suffering from septicemia. A better choice then would be a pure Sulfa product or other medication combination.

Trimethoprim-Sulphamethoxazole/Co-Trimoxazole

The combination of Trimethoprim with Sulphamethoxazole, called co-trimoxazole is widely used in therapeutics in human and veterinary practice. The co-trimoxazole usually contains 5 parts of sulphamethoxazole and 1 part of trimethoprim, In

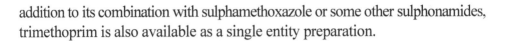

addition to its combination with sulphamethoxazole or some other sulphonamides, trimethoprim is also available as a single entity preparation.

Antimicrobial spectrum

The antibacterial spectrum of co-trimoxazole is broad and includes many Gram-positive and Gram-negative organisms. Gram-positive bacteria that are generally susceptible include most Streptococci and many strains of Staphylococcus and Nocardia. Many Gram-negative organisms of family Enterobacteriaceae are susceptible, but *Pseudomonas aeruginosa* is resistant. Protozoa like Coccidia and Toxoplasma are also affected by the combination. Resistance of co-trimoxazole develops slowly than to sulphamethoxazole and trimethoprim when used alone. In susceptible organisms, resistance usually develops due to acquisition of a plasmid that code for an altered dihydrofolate reductase.

Contraindications

Trimethoprim can lower blood platelets in fish which means any use with fish suffering from any bleeding issues can be often fatally worse since these platelets ("thrombocytes") are blood cells whose function is to stop bleeding.

Pharmacokinetics

Absorption of co-trimoxazole is rapid with peak blood levels occurring between 1-4 hours after oral administration. Trimethoprim being a lipid soluble drug, diffuses well in body tissues and tends to accumulate in acidic environment like that exist in urine. Trimethoprim is metabolized by oxidation followed by conjugation. Both parent drugs and their metabolities are excreted mainly in urine.

Side effects/Adverse effects

Side/adverse effects of co-trimoxazole are the same as when two drugs are used separately. Trimethoprim tends to increase GI and haemolytic effects of sulphonamides; chances of skin toxicity are also more with the combination than the sulphonamide alone. In prolonged therapy, co-trimoxazole may produce folate deficiency because its intestinal synthesis is interfered with existing folate deficiency, chances of toxicity with combination therapy are more in all mammals.

Trimethoprim-Sulphadiazine (Co-Trimazine)

Co-trimazine is a combination of trimethoprim with sulphadiazine in proportions by weight of 1 part to 5 parts respectively. Its pharmacological properties, adverse effects, and clinical uses are similar to that of co-trimoxazole. The normally recommended dose regimen is 30 mg/kg/day for 7-10 days.

Trimethoprim-Sulphadoxine

Trimethoprim-sulphadoxine combination provides a longer duration of action. It is used occasionally for the treatment of sulphadoxine-trimethoprim sensitive infections in domestic animals.

Ormethoprim-Sulphadimethoxine

Ormethoprim [2, 4-diamono-5-(4, 5-dimethoxy-2-methyl benzyl) pyrimidine] is a newly introduced 2, 4 - diaminopyrimidine derivative. It is structurally related to trimethoprim and is mostly used in combination with sulphadimethoxine (1:5). The ormetoprim-sulphadimethoxine shares the mechanism of action and probably bacterial spectrum of activity with co-trimoxazole. The side/adverse effects, contraindications, and drug interactions are also similar to those of co-trimoxazole. It is reported to have comparatively longer duration of action than trimethoprim-sulphamethoxazole combination as its therapeutic levels are well maintained over 24 hours at recommended dosages. The normally recommended dose regimen is 50 mg/kg/day.

Potentiated sulphonamides used for treating fish

Antibiotic	Route of administration	Dosage	Withdrawl time
Trimethoprim-Sulphadiazine	Oral	50 mg/kg	3-4 weeks
Trimethoprim-sulphamethoxazole			
Ormethoprim-sulphadimethoxine			

Uses

Potentiated sulphonamides are active against a wide range of bacterial infections. They have been reported to be active in infections of fish with *Aeromoans salmonicida, A. hydrophila, A. liquefaciens, A. punctata, Vibrio anguillarium, Pasteruella piscicida, Yersinia ruckeri, Edwardsiella ictaluri, E. tarda.* They have shown poor activity against streptococci and are not effective against *Pseudomonas* sp.

9

PROTEIN SYNTHESIS INHIBITORS

A protein synthesis inhibitor is a substance that stops or slows the growth or proliferation of cells by disrupting the processes that lead directly to the generation of new proteins. The substances take advantage of the major differences between prokaryotic and eukaryotic ribosome structures which differ in their size, sequence, structure, and the ratio of protein to RNA. The differences in structure allow some antibiotics to kill bacteria by inhibiting their ribosomes, while leaving human ribosomes unaffected. Translation in prokaryotes involves the assembly of the components of the translation system which are: the two ribosomal subunits (the large 50S & small 30S subunits), the mRNA to be translated, the first aminoacyl tRNA, GTP (as a source of energy), and three initiation factors that help the assembly of the initiation complex. The ribosome has three sites: the A site, the P site, and the E site. The A site is the point of entry for the aminoacyl tRNA. The P site is where the peptidyl tRNA is formed in the ribosome. The E site which is the exit site of the now uncharged tRNA after it gives its amino acid to the growing peptide chain.

In general, protein synthesis inhibitors work at different stages of prokaryotic mRNA translation into proteins like initiation, elongation (including aminoacyl tRNA entry, proofreading, peptidyl transfer, and ribosomal translocation), and termination. The following is a list of common antibacterial drugs and the stages which they target.

Protein synthesis

Attachment of mRNA to 30S subunit

50S binds to 30S to constitute 70S

This unit moves along mRNA so that successive codons of mRNA pass along ribosome from the acceptor site to peptidyl site

A tRNA with its existing aminoacid chain is already attached at the P-site of the complex by complementary codon:anticodon pairing

↓

The incoming tRNA with a new aminoacid binds to the acceptor site by complementary base pairing

Peptide chain on the tRNA attached to the P-site is then transferred to the tRNA linked to A site. This process is called transpeptidation

tRNA which has lost its peptide chain is ejected out from P-site

tRNA at the A-site is translocated to P-site

↓

a freed A-site is now ready to receive new tRNA

AMINOGLYCOSIDES

Aminoglycosides (Aminocylitols) are a group of natural and semi-synthetic antibiotics having aminosugars linked to an aminocyclitol ring by glycosidic bond. They are mostly bactericidal drugs that share many chemical and pharmacological properties and have similar antibacterial spectrum and toxic effects. Important members of the group include Streptomycin, Neomycin, Kanamycin, Gentamicin, Amikacin, and Tobramycin.

History

Streptomycin was the first member of aminoglycoside antibiotics discovered in 1944 by Waksman and co-workers from a strain of *Streptomyces griseus*. Neomycin was next to be isolated in 1949 followed by Kanamycin is 1957 and Gentamicin in 1963. Amikacin was the first semi-synthetic aminoglycoside obtained by chemical modification of Kanamycin. Now aminoglycosides have many members, some of which are extensively used in veterinary medicine.

Chemistry and source

The aminoglycosides consist of two or more amino sugars joined in glycosidic linkages to a hexose (aminocyclitol) nucleus. In streptomycin, the hexose molecule is 2-deoxystreptamine. The presence of amino group in the structure imparts basic nature to aminoglycosides and the hydroxyl group on the sugars provides high water solubility (or poor lipid solubility) to the drugs. If these hydroxyl groups are removed (e.g., tobramycin), the drug becomes more active. Because the groups can be substituted at more than one position on the molecule, several forms of same aminoglycoside may be obtained. For examples, neomycin is a mixture of neomycin B, C, and fradiomycin and gentamicin is a complex of gentamicins C1, C1a, and C2. Minor differences in the chemical structures of these drugs may lead to differences in efficacy and toxicity.

All aminoglycosides are produced by the soil actinomycetes. While most aminoglycosides are obtained by natural fermentation of various species of Streptomyces, some members of the group (e.g., gentamicin) are prepared from

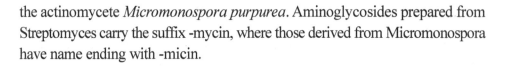

the actinomycete *Micromonospora purpurea*. Aminoglycosides prepared from Streptomyces carry the suffix -mycin, where those derived from Micromonospora have name ending with -micin.

Properties

Members of the aminoglycoside group share some common properties.

a) They are water soluble and polar compounds and generally ionise in solution.

b) They are not absorbed orally, distribute only extracellularly, undergo limited biotransformation and are excreted mainly unchanged in urine.

c) They are bactericidal in action and are more active against gram-negative bacilli, although newer agents have broad-spectrum of activity.

d) They act by interfering with protein synthesis in susceptible bacteria.

e) They are mainly used as sulphate salts, which are highly water soluble. Aqueous solutions of aminoglycosides are stable for months.

f) They are more active in alkaline pH.

g) They have a tendency to cause nephrotoxicity.

h) They have relatively narrow margin of safety.

i) Bacterial resistance to many aminoglycosides develop rapidly. However, an organism resistant to one aminoglycoside may still respond to the other (partial resistance).

j) They show synergistic antibacterial effect with beta-lactam antibiotics, but at high concentrations they can complex with beta-lactam antibiotics resulting in loss of activity.

Classification

On the basis of antibacterial spectrum, aminoglycosides may be classified into following groups:

I. Narrow-spectrum aminoglycosides

e.g., Streptomycin and dihydrostreptomycin.

II. Broad-spectrum aminoglycosides

e.g., Neomycin, Framycetin, Kanamycin and Paromomycin.

III. Extended spectrum aminologlycosides

e.g., Gentamicin, Amikacin, Tobramycin, Sisomicin and Netilmicin.

Mechanism of action

The aminoglycosides are bactericidal drugs that act by inhibiting protein synthesis in susceptible bacteria, mainly gram-negative organisms. Their bactericidal action is concentration dependent i.e., they produce greater cidal effect at high oxygen tension in the environment. The action of aminoglycosides may be broadly divided in two processes – entry of aminoglycosides into bacterial cells and binding of aminoglycosides to bacterial ribosomes.

1. **Passage of aminoglycosides into bacterial cell**: Entry of aminoglycosides into the bacterial cell is a complex process hich which involves:

    a) **Passage across the cell wall**: In this step, aminoglycosides diffuse through the outer coat of Gram-negative bacteria to reach the periplasm. This occurs in a concentration-dependent manner *via* aqueous channels formed by porin proteins (porin channels).

    b) **Passage across the cell membrane**: In the step, aminoglycosides are carried from the periplasmic space into the bacterial cytoplasm. This transportation of aminoglycosides across the cytoplasmic membrane occurs with the help of an oxygen-dependent process, which is linked to the electron transport chain. The oxygen-dependent transport mechanism occurs in two phases – phase I and phase II. In phase I, the ionised aminoglycosides interact with the anionic aminoglycosides with the anionic component of cell membrane (cytoplasmic membrane negatively charged with respect to periplasm and external environment) in a concentration dependent manner. In the next phase (phase II), aminoglycosides are transported actively across the cell membrane into the cytoplasm by a carrier-mediated system. This penetration of aminoglycosides across the cell membrane is favoured by high pH; some aminoglycosides are many times more active in alkaline pH than in acidic medium. The anaerobic environment, in general, inhibits the oxygen-specific carrier transport system. Therefore, anaerobes and facultative anaeorbes are inherently resistant to the action of aminoglycosides.

2. **Binding of aminoglycosides to bacterial ribosomes**: Once inside the bacterial cell, aminoglycosides interact with bacterial ribosomes and inhibit protein synthesis. Aminoglycosides can bind with both 30S and 50S ribosomal subunits, although their binding with 30S subunit is stronger. By binding to ribosomes, aminoglycosides appear to affect a number of steps in protein synthesis, which include:

a) Interference with formation of the initiation complex of protein synthesis.

b) Distortion of m-RNA codon resulting in misreading of the codon. This causes incorporation of one or more incorrect amino acid(s) into the peptide chain and synthesis of abnormal proteins.

c) Promotion of premature termination of translation with detachment of the ribosomal complex (polysomes to monosomes). This leads to incompletely synthesised proteins.

The aminoglycosides vary in the affinity and degree of binding with the ribosomes. They also differ in affecting various steps in protein synthesis. Binding of aminoglycosides to 30S – 50S juncture is probably responsible for their bactericidal effect, in contrast to bacteriostatic drugs that bind to single ribosomal subunit. The cidal action of aminoglycosides also appears to be based on their ability to bring secondary changes in the integrity of cell membrane. These changes are induced during the energy-dependent transportation of aminoglycosides across cell membrane and may occur due to incorporation of defective proteins in the membrane. These alterations result in a membrane that is more permeable to ions, amino acids and even proteins that leak out followed by bacterial death. Altered permeability in cell membrane also leads to augmentation of the carrier-mediated entry of the aminoglycoside antibiotics that reinforce the lethal action.

The important antimicrobial features of aminoglycosides are:

i) The bactericidal action of aminoglycosides is concentration dependent, so they should be given in a single large dose instead of divided doses.

ii) The aminoglycosides are more effective against rapidly multiplying bacteria.

iii) Aminoglycosides show synergism with beta-lactam antibiotics because cell wall injury caused by beta-lactam compounds facilitates increased uptake of aminoglycosides by the bacteria and, thus, their easier accessibility to the bacterial cell membrane.

iv) Some aminoglycosides are transported more efficiently across the bacterial cell membrane than others and, thus, they tend to have grater antibacterial effect.

Biphasic mode of killing

Killing is biphasic, initial rapid killing followed by late slow killing. Those that survive initial killing become resistant to late phase because of decrease in permeability.

Antibacterial spectrum

Antibacterial spectrum of aminoglycosides varies with the type of antibiotic. Streptomycin and dihydrostreptomycin have relatively narrow-spectra, mainly gram-negative species. The broad-spectrum aminoglycosides (e.g., neomycin and kanamycin) are active against many Gram-negative and Gram-positive organisms, but not *Pseudomonas*. Anaerobic bacteria are only moderately sensitive to aminoglycosides.

Bacterial resistance

The susceptible microorganisms may acquire resistance to aminoglycosides by more than one mechanism. These may be plasmid mediated or acquired by mutation.

1. **Inactivating enzymes**: The susceptible microorganisms acquire resistance to aminoglycosides mainly by producing transferase enzymes like aminoglycosyl acetyltrasferase, aminoglycosyl nucleotidyl-transferase, and aminoglycosyl phosphotransferase in the periplasm. These transferases catalyse the conjugation of specific hydroxyl or amino groups on aminoglycosides with acetyl, adenyl, and phosphorous groups to inactivate the antibiotics. The conjugated aminoglycosides then compete with the parent drug for transport in periplasm and once in cytoplasm they fail to bind to ribosomes. The conjugated aminoglycosides are also incapable of enhancing the active transport like the unaltered drug. These inactivating enzymes are acquired by susceptible bacteria mainly by conjugation and transfer of plasmids. The susceptibility of aminoglycosides to specific transferase enzymes is quite variable, which provides differences in their susceptibility patterns. Some aminoglycosides such as amikacin are relatively resistant to these inactivating enzymes due to chemical modifications.

2. **Impaired transport**: Impaired transport of aminoglycosides across the cell wall / membrane may provide resistance to susceptible microorganisms. This may occur due to changes either in the pores (porin size in the cell wall) that becomes less permeable to antibiotics or in the active transport system in the cell membrane.

3. **Alteration in ribosomal structure**: Alterations in the ribosomal structure can confer high degree resistance by decreasing the binding of aminoglycosides with their target sites.

Pharmacokinetics

The pharmacokinetic features of the aminoglycosides in fish are similar to most veterinary species.

Absorption: The aminoglycosides are poorly absorbed (<10%) from the gastrointestinal tract; therefore this route is reserved for the treatment of GI infections. The absorption from IM injection site is rapid and nearly complete (>90%) and peak levels in blood are achieved within 30 to 45 minutes of administration.

Distribution: Aminoglycosides are extensively distributed in extracellular fluid, but they do not readily enter into cells due to their polar nature. The aminoglycosides bind poorly (<20%) to plasma proteins. The volume of distribution of aminoglycosides is directly proportional to the volume of extracellular fluid.

Biotransformation and excretion: The aminoglycosides that are not metabolized in body are excreted largely as such (~90%) in urine by glomerular filtration. Plasma half-lives of aminoglycosides is similar and varies between 2 to 4 hours in animals / patients with normal renal function. In renal insufficiency, half-lives may increase to 24-48 hours. The disposition of aminoglycosides varies among animals because of differences in glomerular filtration rates. The clearance also depends on the volume of distribution.

Side effects/Adverse effects

All aminoglycosides have potential to produce toxic effects, but the relative propensity differs. Nephrotoxicity and neuromuscular blockade are important adverse effects observed with aminoglycosides.

1. **Nephrotoxicity**: Nephrotoxicity with aminoglycosides occurs as a result of excessive accumulation (40 to 50 times the levels in blood) of antibiotics by the proximal tubular cells in kidneys. As aminoglycosides are positively charged agents, they get attracted to negatively charged phospholipids of the renal membrane followed by their transport inside the tubular cells *via* pinocytosis. In renal tubules, the aminoglycosides get sequestered into lysosomes and also interact with cell organelles resulting in cellular death. The aminoglycosides can inhibit various essential enzymes like phospholipases, sphingomyelinases, and ATPases. Inhibition of phospholipase, in particular, results in reduced synthesis of prostaglandins, diacyl glycerol and inositol triphosphate. As prostaglandins play important role in regulation of renal blood flow, their reduced synthesis has direct effect on glomerular filtration.

2. **Neuromuscular blockade**: All aminoglycosides have potential to produce neuromuscular blockade. The effect is produced mainly by interference with acetylcholine release from motor nerve endings, probably by antagonism of

Ca++ that is normally required for exocytosis. Neuromuscular blockade usually occurs after IV or IP administration of aminoglycosides, but can also be seen with other routes of drug administration

Contraindications and precautions

Aminoglycosides are contraindicated in animals that are hypersensitive to them. They should be used with extreme caution in pre-existing renal disease and in neonatal animals. In general, aminoglycosides should be reserved for severe Gram-negative infections and should be used only when the benefits of therapy exceed the potential risks.

Drug interactions

Aminoglycosides show interactions with drugs that cause nephrotoxicity and neurotoxicity. Aminoglycosides show synergistic antibacterial effect when used with β-lactam antibiotics. However, the concurrent use of aminoglycosides with cephalosporins is controversial due to additive nephrotoxic effect. High concentrations of Carbenicillin, Ticarcillin, and Piperacillin are reported to inactive aminoglycosides in presence of renal failure.

1. NARROW-SPECTRUM AMINOGLYCOSIDES

Streptomycin

Streptomycin is the oldest aminoglycoside obtained from *Streptomyces griseus*. It occurs as a white crystalline substance, which is water-soluble. Its use has declined in veterinary practice, except in combination with penicillins. Its mechanism of action is typical of aminoglycosides. Antibacterial spectrum of streptomycin is relatively narrow. Some organisms susceptible to streptomycin are *E.coli, Salmonella, Klebsiella, Pasteurella and Mycobacterium tuberculosis*.

Bacterial resistance to streptomycin develops rapidly. However, only partial and often unidirectional cross-resistance occurs between streptomycin and other aminoglycosides.

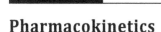

Pharmacokinetics

Pharmacokinetics of streptomycin is typical of aminoglycosides. It is highly ionised antibiotic, so is not absorbed from GI tract. After IM administration, streptomycin is gradually absorbed from the injection site with peak plasma concentration achieved in about 60-90 minutes after administration.

Dihydrostreptomycin

Dihydrostreptomycin is related structurally and pharmacologically to Streptomycin.

1. BROAD-SPECTRUM AMINOGLYCOSIDES

Neomycin

Neomycin is a broad-spectrum amino-glycoside obtained from *Streptomyces fradiae*. Neomycin is actually a complex of three compounds-neomycin A (neamine), neomycin B (framycetin) and neomycin C. Neomycin A is biologically inactive, whereas neomycin B and C are biologically active. The commercially available product of neomycin almost entirely consists of the sulphate salt of neomycin B. Neomycin sulphate exists as an odourless, white to slightly yellow hygroscopic powder that is freely soluble in water, slightly soluble in alcohol, and practically insoluble in acetone and chloroform. Neomycin is a highly stable antibiotic.

Mechanism of action

The mechanism of action of neomycin is typical of aminoglycosides.

Antibacterial spectrum

Neomycin has a broad antibacterial spectrum and includes many Gram-positive aerobic bacteria as well as several Gram-positive organisms. However, *Pseudomonas aeruginosa* and *Streptococcus pyogenes* are not sensitive.

Pharmacokinetics

Neomycin is poorly (~3%) absorbed following oral administration. After IM injection, therapeutic levels can be achieved gradually with peak plasma levels occurring in 1 hour of dosing. Similar to other aminoglycosides, it is negligibly metabolized and excreted largely as unchanged drug in the urine. Orally administered neomycin also remains unmetabolised in GI tract and is excreted unchanged in the faeces.

Framycetin/Neomycin B

Framycetin is the term used to describe the neomycin B when present without other components. The properties of framycetin are essentially similar to those of neomycin discussed above. Similar to neomycin, framycetin sulphate is used by oral and topical routes for the treatment of enteritis and ear/eye infections, respectively.

Kanamycin

Kanamycin is an aminoglycoside antibiotic produced by *Streptomyces kanamyceticus*. It consists of three components – Kanamycin A, the major component (usually designated as Kanamycin) and Kanamycin's B and C, two minor components. Kanamycin is chemically and pharmacologically related to neomycin. It once demonstrated considerable activity against most aerobic Gram-negative bacilli, but now its use has declined due to development of resistance. Presently, it is used mainly for topical application and by oral route for enteric infections.

Paromomycin

Paromomycin (Aminosidine) is an antibiotic isolated from various *Streptomyces* organisms such as *S. rimosus var. paromomycinus, S. catenulae, or S. chrestomyceticus*. It differs from neomycin and framycetin in that it possesses wide spectrum of activity including both Gram-negative and Gram-positive bacteria, tapeworms and some GI protozoan like *Entamoeba histolytica and*

Giardia. Similar to other aminoglycosides, paromomycin is poorly absorbed from the GI tract. Its use in veterinary medicine is limited to treatment of enteric infection caused by susceptible microorganisms.

III. EXTENDED-SPECTRUM AMINOGLYCOSIDES

Gentamicin

Gentamicin is the most widely used aminoglycoside with extended-spectrum of antibacterial activity. Gentamicin is obtained from *Micromonospora purpurea*; the spelling ending in-micin is to indicate that the source is not Streptomyces. Gentamicin exists as white amorphous powder that is freely soluble in water, moderately soluble in ethanol and acetone, and practically insoluble in benzene.

Mechanism of action

The mechanism of action of Gentamicin is typical of aminoglycosides discussed previously.

Antibacterial spectrum

Gentamicin is a broad-spectrum aminoglycoside with extended spectrum. It is active against a wide-range of Gram-negative and Gram-positive bacteria including *Staphylococcus, Streptococcus, Pseudomonas, Proteus, E.coli, Klebsiella* species. Its activity against anaerobes is weak. Organisms that are resistant to gentamicin are generally also resistant to streptomycin and neomycin. However, many bacteria resistant to other aminoglycosides are sensitive to Gentamicin.

Pharmacokinetics

Gentamicin, similar to other aminoglycosides, is poorly absorbed from GI tract after oral administration. Following IM injection, peak plasma concentration is attained in 30-40 minutes. Subcutaneous administration results in slightly delayed peak levels. Bioavailability following IM administration is greater than 90%. Binding to plasma proteins is low. It is excreted mainly unchanged in the urine.

The elimination half-life of gentamicin has been reported to be between 1 to 3 hours in various species. Animals with decreased renal functions can have significantly prolonged half-life.

Side effects/Adverse effects

Gentamicin has potential to cause nephrotoxicity, ototoxicity and neuromuscular blockade in the same way as other aminolglycosides. However, it is considered comparatively less toxic than many aminoglycosides. Its contraindications and drug-interactions are similar to other

Amikacin

Amikacin is a semi-synthetic aminoglycoside and an acetylated derivative of kanamycin A. It occurs as white, crystalline powder that is sparingly soluble in water. The spectrum of amikacin is broadest of the aminoglycoside group with activity against a wide range of microorganisms including some that are resistant to gentamicin and tobramycin. These organisms include *E.coli* and most species of *Klebsiella, Proteus, Pseudomonas, Salmonella, Enterobacter, Serratia, Shigella, Mycoplasma, and Staphylococcus*. Amikacin has unique position in the group because it is resistant to aminoglycoside inactivating enzymes. Amikacin like other aminoglycosides is poorly absorbed by oral route. Elimination of amikacin following parenteral administration occurs entirely by glomerular filtration.

Tobramycin

Tobramycin is an aminoglycoside obtained from *Streptomyces tenebrarius*. It is structurally related to kanamycin but has improved antibacterial activity similar to gentamicin. Gentamicin resistant organisms also respond to tobramycin, particularly if resistance is due to aminoglycoside degrading enzymes. The pharmacokinetic properties and adverse effects of tobramycin are very similar to that of gentamicin. Tobramycin may be used clinically to treat serious Gram-negative infections, particularly those resistant to gentamicin.

Sisomicin and Netilmicin

Sisomicin and netilmicin are extended spectrum aminoglycoside antibiotics used primarily in human practice. Sisomicin is a natural aminoglycoside obtained from *Micromonospora inyoensis* and is chemically and pharmacologically similar to gentamicin. Netilmicin is a semi-synthetic derivative of sisomicin, which has broader spectrum of activity than gentamicin and is relatively resistant to aminoglycoside inactivating enzymes. These drugs are presently undergoing clinical trials for use in veterinary practice.

Other Aminoglycoside Antibiotics

Several other aminoglycoside antibiotics with variable antibacterial activities have been used in human medicine. These include butikacin, butirocin, fortimicin, lividomicin, propikacin, ribostamycin, sagamycin, seldomycin and sorbistin.

TETRACYCLINES

Tetracylines are a group of broad-spectrum antibiotics having a nucleus of four cyclic rings. They are obtained either naturally from soil actinomycetes or prepared semi-synthetically. They have similar antimicrobial features, but differ somewhat from one another in terms of their spectra and pharmacokinetics.

History

Tetracycline antibiotics were produced by systemic screening of soil microorganisms. The first member of the group was chlortetracycline derived from soil actinomycete *Streptomyces aureofaciens* introduced in 1948. This was followed by oxytetracycline. Removal of chlorine atom from chlortetracycline produced semi-synthetic tetracycline was introduced in 1952. Further discovery led to other semi-synthetic tetracycline like methacycline, doxycycline, and rolitetracycline. Doxycycline and minocycline are newer tetracyclines with high lipid solubility and longer duration of action.

Classification

Tetracyclines are classified according to their duration of action.

I.   Short-acting tetracyclines (t1/2=<8 hours)

e.g., oxytetracycline, tetracycline and chlortetracycline

II. Intermediate acting tetracyclines (t1/2=8-16 hours)

e.g., demeclocycline and methacycline.

III. Long-acting tetracyclines (t1/2>16 hours)

e.g., doxycycline and minocycline.

Mechanism of action

Tetracyclines inhibit bacterial protein synthesis and are primarily bacteriostatic. In a way somewhat similar to aminoglycosides, the action of tetracyclines can be divided into two processes-passage of tetracyclines into bacterial cell and interaction of tetracyclines with bacterial ribosomes.

1. **Passage of tetracyclines into bacterial cell**: Tetracyclines enter gram-negative bacteria by two transport mechanisms: in part by a passive process and in part by an active process. The first is passive diffusion through the hydrophilic channels formed by the porin proteins in outer cell membrane. The more lipid soluble members (e.g., doxycycline and minocycline) pass directly through the lipid bilayer by passive diffusion. The second mechanism involves an energy dependent active transport system that pumps all tetracyclines across cytoplasmic membrane. Although passage of tetracyclines into gram-positive bacteria is less well understood, it requires an energy-dependent carrier transport mechanism.

2. **Interaction of tetracyclines with bacterial ribosomes**: Once the tetracyclines gain access to bacterial cell, they bind to the 30 S ribosomal subunit. They prevent binding access of aminoacyl t-RNA-ribosome complex. This prevents addition of amino acids to the growing peptide chain resulting in inhibition of protein synthesis. Tetracyclines are more effective against multiplying microorganisms and are more active at pH 6-6.5. Effects of

tetracyclines are mostly reversible as the bacterial protein synthesis is restored when the drug is removed. Therefore, a responsive host-defense system is essentially required to remove static bacteria. Although the tetracyclines are primarily bacteriostatic, at high concentrations they tend to become bactericidal because at high concentrations they appear to affect the functional integrity of bacterial cell membranes as well. At high concentrations, they also tend to impair protein synthesis in host cells by binding to the eukaryotic 40S ribosomal subunit. However, penetration of tetracyclines in eukaryotic cells through cell membrane is poor due to lacking of specific carrier transport system. Moreover, the mammalian protein synthesizing apparatus is less sensitive to tetracyclines.

Antimicrobial spectrum

The tetracyclines are broad-spectrum antibiotics. They are active against a wide range of aerobic and anaerobic Gram-positive and Gram-negative bacteria. Tetracyclines are ineffective against fungi and viruses. Oxytetrcycline has been used as a first choice drug for nearly all bacterial diseases of fish. It can be used for Cold water vibriosis (*Vibrio salmonicida*), Enteric redmouth (*Yersinia ruckeri*), Furunculosis (*Aeromonas salmonicida*) and, Columnaris (*Flexibacter columnaris*).

Microbial resistance

Microbial resistance to tetracyclines develops slowly in a graded manner. Microbial resistances to tetracyclines may be acquired by 3 different mechanisms. The most important mechanism is decreased penetration of the drug into previously sensitive microorganisms. This may occur either due to decreased antibiotic influx into bacterial cell as a result of development of less efficient transport mechanisms or due to an energy dependent efflux of antibiotic from bacterial cell by a modified protein carrier. The enzymatic inactivation of tetracyclines and production of proteins by microorganisms that protect ribosomes by binding with tetracyclines are other two lesser important mechanisms of acquiring resistance.

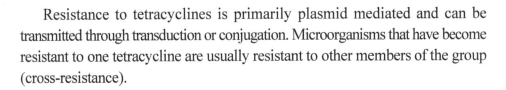

Resistance to tetracyclines is primarily plasmid mediated and can be transmitted through transduction or conjugation. Microorganisms that have become resistant to one tetracycline are usually resistant to other members of the group (cross-resistance).

Pharmacokinetics

Absorption: Absorption of tetracyclines from GI tract is decreased in presence of polyvalent cations (e.g., Ca^{++}, Mg^{++}, and Fe^{+++}) that are present in food. All tetracyclines produce varying degree of tissue irritation on parenteral administration, especially chlortetracycline. Therefore for parenteral administration, buffered solutions are prepared. Procaine is added to tetracyclines solution for IM administration in case of small animals.

Distribution: Tetracyclines bind to plasma proteins to varying degrees and are widely distributed in most tissues including kidneys, liver, lungs, bile, and bones. **Biotransformation and excretion:** With exception of lipid soluble tetracyclines, the tetracycline antibiotics are not metabolized to a significant extent in the body. Most tetracyclines are excreted in urine (~60%) *via* glomerular filtration pathway and in faeces (~40%) *via* biliary excretion. Tetracyclines undergo enterohepatic circulation, which may affect their duration of action.

Side effects/Adverse effects

The tetracyclines have a relatively low toxicity at normal dosage levels. However, a number of adverse effects have been associated with tetracyclines. Adverse effects may be worsened in animals with renal disease due to decreased elimination of the drug.

Gastrointestinal upsets: All tetracyclines produce GI irritation to varying degree in some patients, particularly after oral administration.

Effect on bones/teeth: Tetracyclines are deposited in growing teeth and bones due to their chelating properties with calcium. They form tetracycline-calcium orthophosphate complex, which inhibits calcification of the teeth.

Nephrotoxicity: Tetracyclines are potentially nephrotoxic, particularly in renal insufficiency.

I. SHORT-ACTING TETRACYCLINES

Oxytetracycline

Oxytetracycline (Terramycin) is the most commonly used tetracycline in veterinary practice today. It is obtained from the actinomycete *Streptomyces rimosus*. Oxytetracycline base occurs as a pale yellow to tan, crystalline powder that is very slightly soluble in water and sparingly soluble in alcohol; oxytetracycline HCl is readily soluble in alcohol. Commercially available oxytetracycline HCl injections are usually prepared in either propylene glycol or povidone based products. It shares the antibacterial spectrum and mechanism of action with other tetracyclines.

Oxytetracycline is readily absorbed after oral administration with bioavailability of about 60-80%. After IM administration, its peak plasma levels may be achieved in 0.5-2 hours; After absorption, it is widely distributed in body tissues and fluids. Excretion of oxytetracycline is mainly as unchanged drug in urine and bile. Oxytetracycline is widely used in veterinary species for treatment of various infections caused by susceptible organisms.

Tetracycline

Tetracycline is a broad-spectrum antibiotic obtained naturally from *Streptomyces aureofaciens* or derived semi-synthetically from oxytetracycline. It is slightly less soluble than oxytetracycline, but is more stable in solution. Tetracycline HCl is more stable in water. Its mechanism of action and antimicrobial spectrum are similar to that of oxytetracycline. Like oxytetracycline, it is readily absorbed (60-80%) after oral administration. However after IM injection, its absorption is less. Distribution, metabolism, and excretion of tetracycline are similar to those of oxytetracycline.

Chlortetracycline

Chlortetracycline (Aureomycin) was the first antibiotic of the tetracycline group isolated from the substrate of *Streptomyces aureofaciens*. It occurs as golden-yellow crystals that are slightly soluble in water. The chlortetracycline HCl is fairly soluble in water but has a strong bitter taste. Chlortetracycline has antimicrobial spectrum and pharmacology almost similar to oxytetracycline. However, it is absorbed only to the extent of 30% after oral administration. Intramuscular injection is painful and is not recommended.

II. Intermediate-acting tetracyclines

Demeclocycline

Demeclocycline (Demethylchlortet-racycline; ledermycin) is an intermediate acting tetracycline. Demeclocycline is chlortetracycline without the methyl group. It is more soluble and more active than the chlortetracycline. After oral administration, demeclocycline is rapidly absorbed and distributed in body, but its excretion is slower that gives longer effective blood levels than observed with the oxytetracycline or chlortetracycline.

Methacycline

Methacycline (Rondomycin) is a semi-synthetic tetracycline prepared from oxytetracycline. Its efficacy is comparable with oxytetracycline, but is possesses longer duration of action (t1/2=14-16 hours). It is less used in veterinary practice.

III. Long-acting tetracyclines/newer tetracyclines

Doxycycline

Doxycycline is a semi-synthetic tetracycline derived from oxytetracycline or methacycline. It is more lipophilic than the older tetracyclines and has a number of advantages including longer duration of action.

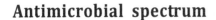

Antimicrobial spectrum

Doxycycline has an antimicrobial spectrum very similar to oxytetracycline, but some strains of bacteria may be more susceptible to doxycycline due to its better penetration into the microbial cells.

Pharmacokinetics

Doxycycline is rapidly and nearly completely (90-100%) absorbed from GI tract after oral administration. Doxycycline penetrates better into several body compartments. It also shows greater plasma protein binding (~90%) than older tetracyclines, which prolongs its elimination half-life. Unlike other tetracyclines, doxycycline is significantly metabolized (up to 40%) in the body and excreted mainly as inactive metabolites (chelates or conjugates) in the faeces. In these inactive forms, the antibiotic has insignificant effect on the lower intestinal microflora.

Minocycline

Minocycline is a semi-synthetic derivative of oxytetracycline. Similar to doxycycline, it is a lipid-soluble tetracycline that result in better antimicrobial properties, efficient oral absorption (~95-100%), higher tissue penetration, greater plasma protein binding and longer elimination half-life. Like doxycycline, minocycline is excreted mainly in faeces as inactive metabolites and parent drug.

Clinical uses of Tetracyclines

Oxytetracycline will treat Gram-positive (such as Streptococcus) and some Gram-negative bacteria (some *Vibrio* sp.) found in aquarium/pond environments. Maladies include, body slime and cloudy eyes (its best use), marine ulcer disease, cold water disease, bacterial hemorrhagic septicemia redness in the body, open sores or loss of scales and mouth fungus.

DOSAGE: 250- 500 mg per 20 gallons every 48 hours (24 hours for severe problems) with a 50% water change before each treatment. This antibiotic is best used mixed in with food, especially if your pH is above 8.0.

Contraindications

- Best not used concurrently with other antibiotics or chemical treatments, although use in a bath with Methylene blue is effective.

- Do not use with fish suffering from ammonia/nitrite poisoning or with fish with low red blood cell count (anemia) such as suffering from an acute gill infestation of Oodinium, Flukes or similar parasites that attacks the gills and thus renders a fish anemic.

- As with all Tetracycline products, less effective in higher calcium environments

- Do not use with any iron oxide containing products

AMPHENICOLS

Amphenicols are a group of broad-spectrum bacteriostatic drugs that include chloramphenicol, the parent compounds, and its congener's thiamphenicol and florfenicol.

Chloramphenicol

Chloramphenicol is a broad-spectrum bacteriostatic antibiotic. Chloramphenicol was initially obtained from *Streptomyces venezuelae* in 1947, but now is manufactured synthetically. Chloramphenicol is used in a variety of infections in veterinary medicine, particularly those caused by anaerobic bacteria.

Properties

Chloramphenicol is a highly lipid soluble antibiotic that exists as a yellowish white crystalline solid. It is freely soluble is alcohol and acetone, fairly soluble in ether, less soluble in water, and insoluble in benzene. Chloramphenicol is marketed either as a free base or in ester forms (e.g., palmitate, succinate, or pantothenate salts). Chloramphenicol palmitate is insoluble in water and sparingly soluble in alcohol, so is used for oral administration; chloramphenicol, succinate (sodium succinate) is freely soluble in both acetone and water and, therefore, is used for parenteral injection. The inert chloramphenicol esters are hydrolysed *in vivo* to free biologically active chloramphenicol.

Chemistry and structure-activity relationship

Chloramphenicol (D-threo-2-dichloro-acetamido-1-p-nitrophenyl-1, 3-propanediol) is a derivative of nitrobenzene and dichloroacetic acid. It is unique among natural compounds having a nitrobenzene moiety in its structure. Although chloramphenicol exists in 4 stereoisomers, only D-threo form has antibacterial activity. The para-nitro group is not important for antibacterial activity. The p-nitro phenyl group may be changed to ortho-or meta-nitro compounds or even be replaced by halogens without significant loss of activity. The p-nitro group has implicated in the irreversible suppression of bone marrow.

Mechanism of action

Chloramphenicol inhibits protein synthesis in susceptible microorganisms and, to a lesser extent, in mammalian cells. Chloramphenicol readily penetrates into bacterial cells, probably both by passive and facilitated diffusions. Once inside the bacterial cell, it binds reversibly to the 50 S ribosomal subunit and prevents the activity of peptidyl transferase enzyme. This interferes with transfer of elongating polypeptide chain in the newly attached aminocyl t-RNA at ribosome-mRNA complex. By binding specifically to 50S ribosomal subunit, chloramphenicol also appears to hinder the access of aminoacyl tRNA to acceptor site of the tRNA but its codon is not affected, but failure of aminoacyl-tRNA to associate properly with the acceptor site prevents the subsequent transpeptidation reaction catalyzed by peptidyl transferase. Although host ribosomes do not bind as effectively as do bacterial ribosomes, some host ribosomal protein synthesis is impaired. The effect of chloramphenicol is usually bacteriostatic, but at high concentrations or against some very susceptible organisms it can be bactericidal.

Antimicrobial spectrum

Chloramphenicol is a broad-spectrum antimicrobial agent. It is active against many Gram-positive and Gram-negative bacteria, and both anaerobes and aerobes including *Staphylococcus, Streptococcus, Salmonella, Brucella,*

Shigella, Neisseria, and Haemophilus species. Many gram-negative anaerobes including *Clostridium, Bacteroides, Fusobacterium, and Veillonella* are also reported to be sensitive to chloramphenicol. Chloramphenicol also has activity against *Nocardia, Rickettsia, Chlamydia, and Mycoplasma*. Chloramphenicol has no/less activity against many strains of *Pseudomonas* and *proteus* and no activity against *Mycobacterium*, protozoa, fungi, and viruses.

Bacterial resistance

Most bacteria are capable of developing resistances to chloramphenicol. Resistance against chloramphenicol develops slowly in a graded manner, as with tetracyclines. Resistance in usually plasmid mediated and occurs mainly due to production of chloramphenicol acetyltransferase enzymes in bacteria that inactivate the drug. In resistant Gram-negative bacteria, chloramphenicol acetyltransferase is a constitutive enzyme, whereas in Gram-positive bacteria it is inducible. Non-enzymatic resistance like decreased permeability of antibiotic into the bacteria (e.g., in *E.coli* and *Pseudomonas*) and lowered affinity of bacterial ribosome for chloramphenicol are other mechanisms of resistance. The bacterial resistance to chloramphenicol often occurs together with resistance to tetracyclines, erythromycin, streptomycin, ampicillin, and plasmids that code to multi-drug resistance.

Pharmacokinetics

Chloramphenicol is a highly lipid-soluble drug, so it is very rapidly and efficiently absorbed after oral administration. The peak plasma concentration is generally achieved within 30 minutes of oral dosing. After absorption, chloramphenicol diffuses throughout the body and reaches sites of infection inaccessible to many other antibacterial drugs. Chloramphenicol binds to plasma proteins to the extent of 30-60%. Chloramphenicol is primarily eliminated by hepatic metabolism *via* glucuronide conjugation. The inactive glucuronide metabolite as well as some active unchanged drug (~15-20%) is excreted mainly in urine by glomerular filtration and tubular secretion. Fishes are deficient in glucuronide conjugation, so elimination of chloramphenicol is slower.

Side effects/Adverse effects

Chloramphenicol has a low order of toxicity in domestic animals when used in recommended doses. However, chloramphenicol may produce dose-dependent (reversible) and dose-independent (irreversible) bone marrow depression. The reversible bone marrow depression may be observed in all species of animals, especially with prolonged use of the antibiotic.

Contraindications and precautions

Chloramphenicol is contraindicated in animals that are hypersensitive to it. The potential of haematopoietic toxicity makes chloramphenicol unsuitable for fishes with pre-existing haematological disorders. Chloramphenicol should be avoided in fish fingerlings because of less developed metabolizing enzymes.

Drug interactions

Macrolide antibiotics that bind to 50 S ribosomal subunit of susceptible bacteria may potentially antagonize the activity of chloramphenicol by competing for the same binding site. Chloramphenicol should not be concurrently used with β-lactam and aminoglycoside antibiotics as bacteriostatic action of former drug may inhibit the bactericidal action of it.

It is used as prophylactic agent against carp dropsy (*Aerobacterium liquefaciens*) and also used in treatment of trout ulcer disease caused by *Haemophilus piscium* and Furunculosis caused by *A. salmonicida.*

Thiamphenicol

Thiamphenicol is a semi-synthetic derivative of chloramphenicol, where p-nitro group is replaced by a sulphometyl group ($CH_3SO_2^-$). This structural modification preserves the antibacterial activity of chloramphenicol, but makes the thiamphenicol less lipid soluble and less potent. The antibacterial spectrum and mechanism of action of thiamphenicol resembles that of chloramphenicol, but the pharmacokinetics is different from that of chloramphenicol. Being less lipid soluble, thiamphenicol diffuses more slowly through lipid membranes and is not metabolized to a significant extent. It is excreted rapidly mainly as unchanged active compound in the urine. Hence, it is considered to be less toxic than the chloramphenicol. Thiamphenicol is used for the treatment of various infections normally susceptible to chloramphenicol.

Florfenicol

Florfenicol is a fluorinated analogue of thiamphenicol in which the hydroxyl group oxide chain is replaced with fluorine. This structural modification makes the antibiotics less susceptible to microbial inactivation and also abolishes the occurrence of irreversible aplastic anaemia. Florfenicol is a broad-spectrum antibiotic with mechanism of action and antibacterial spectrum similar to chloramphenicol. Florfenicol is reported to be active against certain bacteria that are resistant to chloramphenicol, well absorbed after oral and parenteral administration and is safe at recommended doses.

MACROLIDES AND LINCOSAMIDES

MACROLIDES

The macrolides (macrocyclic lactones) are a group of bacteriostatic antibiotics that structurally consist of a large lactone ring attached to deoxy sugars. They include Erythromycin, Oleandomycin, Troleandomycin, Spiramycin, Josamycin, Tilmicosin and Tylosin. Roxithromycin, Clarithromycin, and Azithromycin have been recently introduced.

Chemistry and source

The macrolides or macrocyclic lactones that have complex chemical structures consisting of a large lactone ring, usually 14-16 atoms, attached to deoxy sugars by glyosidic linkages. Each macrolide antibiotic may further consists of a mixture of closely related agents that differs from each other with respect to some chemical substitution in the structure (e.g., erythromycin consists of erythromycins A,B,C,D, and E). Macrolides are mostly obtained from various species of Streptomyces, soil-borne bacteria and some are prepared semi-synthetically.

Mechanism of action

The macrolide antibiotics are usually bacteriostatic in action, but may become bactericidal at high concentrations against susceptible microorganisms. Similar to other drugs inhibiting protein synthesis, the action of macrolides can be divided into two processes-passage of macrolides into bacterial cell and interaction of macrolides with bacterial ribosomes.

1. **Passage of macrolides into bacterial cells**: Macrolides are transported into the cytoplasm of susceptible microorganisms by an active transport system. The Gram-positive bacteria accumulate about 100 times more antibiotics than do Gram-negative organisms. The non-ionised form of the macrocyclic antibiotic is considerably more permeable to bacterial cells, so the drugs show enhanced antimicrobial activity at alkaline pH.

2. **Interaction of macrolides with bacterial ribosome**: Once inside bacterial cells, macrolides bind to the 50 S ribosomal subunit and interfere with the bacterial protein synthesis. The binding site of macrolides on the ribosomes is near but not identical to that of chloramphenicol. It is believed that macrolides do not inhibit peptide bond formation but block mainly translocation step where in a newly synthesized peptidyl-tRNA molecule moves from the acceptor (A site) on ribosome to the peptidyl (P) site. Failure of translocation stops protein synthesis as the A site is not available for the next incoming aminoacyl-tRNA and ribosomal complex cannot move to the next codon. The effect is more pronounced on rapidly dividing bacteria and mycoplasma. Macrolides generally do not bind to mammalian ribosomes.

Antimicrobial spectrum

Macrolides are medium spectrum antibiotics mainly effective against Gram-positive bacteria, *Chlamydia* and *Rickettsia*. Because a majority of fish bacterial pathogens are Gram-negative, the indications are limited and specific; the important ones are Bacterial Kidney Disease (BKD), Streptococcosis, especially in yellowtail.

Bacterial resistance

Acquired resistance to macrolide antibiotics in susceptible Gram-positive bacteria result mainly from changes in ribosomal structure and loss of macrolide affinity. Decreased permeability of antibiotic through cell envelops, efflux of drug by an active pump mechanism and increased production of inactivating enzyme have also been suggested. The bacterial resistance is mostly plasmid-mediated that may develop either rapidly (e.g., erythromycin) or slowly (e.g., tylosin). Cross-resistance with other macrolides, lincosamides, and chloramphenicol may occur due to their similar binding site on the ribosome.

Pharmacokinetics

Absorption: Macrolides are lipid soluble antibiotics that are rapidly absorbed from the GI tract, if not inactivated by gastric acid. Oral preparations are often mixed with feed to prevent gastric acid inactivation.

Distribution: Macrolides are widely distributed in tissues and tend to accumulate in some cells including macrophages because of ion trapping.

Biotransformation and excretion: Macrolide antibiotics undergo extensive biotransformation (~80%) in liver through microsomal enzyme system. They are excreted in both active and inactive metabolite forms mainly in bile (>60%) and undergo enterohepatic cycling.

Drug interactions

Macrolide antibiotics should not be used with chloramphenicol or lincosamides because they compete for the same/similar binding sites on 50S ribosomal subunit. Erythromycin and troleandomycin inhibit microsomal enzymes and depress hepatic oxidation of many drugs like warfarin, theophylline, carbamazepine, and methylprednisolone. Activity of macrolides decreases in acidic environment.

Erythromycin

Erythromycin is the most widely used and prototype macrolide antibiotic. It was isolated from *Streptomyces erythreus* in 1952. Erythromycin is available commercially in several salt and ester forms including erythromycin-estolate (propionate lauryl sulphate), -stearate, -ethylsuccinate, -lacto-bionate, and – gluceptate (glucoheptonate).

Antibacterial spectrum

The antimicrobial spectrum of erythromycin is narrow including mostly gram-positive and a few gram-negative organisms. It is highly effective against gram-positive cocci like *Streptococci and Staphylococci* and bacilli like *Corynebacterium*. Other susceptible organisms are *Campylobacter, Chlamydia, Mycoplasma, Haemophilus, Pasteurella, Brucella, Listeria, Clostridium*. Most strains of Enterobacteriaceae are resistant to erythromycin.

Pharmacokinetics

Erythromycin base is incompletely but adequately absorbed from upper small intestine after ingestion. Erythromycin can be given IM but absorption is slow, pain and swelling occurs at injection site. Bioavailability after IM or SC injection is only about 40-60%. Erythromycin is well distributed throughout the body fluids and tissues. It is 70-80% plasma protein bound, partly metabolized in liver and excreted primarily in the bile in active form. Some of the drug is reabsorbed after biliary excretion (enterohepatic cycling) which prolongs its elimination.

Tylosin

Tylosin is a macrolide antibiotic obtained from a strain of *Streptomyces fradiae*. It is structurally and pharmacologically related to the erythromycin. The antibacterial spectrum and mechanism of action of tylosin is similar to that of erythromycin. It is suggested that in addition to inhibiting translocation steps in bacterial protein synthesis, tylosin may inhibit binding of aminoacyl tRNA. Development of resistance is low in susceptible microorganisms in comparison with erythromycin. It shows cross-resistance to erythromycin and other macrolide antibiotics. The pharmacokinetics of tylosin is generally similar to that of erythromycin.

Tilmicosin

Tilmicosin is a semi-synthetic macrolide antibiotic. Like erythromycin, tilmicosis has bacteriostatic action against mainly Gram-positive bacteria, although some Gram-negative bacteria like *Pasteurella* are also equally affected. It is also reported to have good efficacy against Mycoplasma. Unlike erythromycin and other macrolide antibiotics, tilmicosin is a potentially toxic antibiotic. It produces cardiovascular toxicity in several species, particularly in high doses.

Spiramycin

Spiramycin is obtained from a strain of *Streptomyces ambofaciens*. Its antimicrobial spectrum and mechanism of action are similar to those of erythromycin. Like other macrolides, it is useful in many gram-positive infections resistant to penicillins.

Oleandomycin and Troleandomycin

Oleandomycin and related compound troleandomycin (triacetyl oleandomycin) are less commonly used macrolide antibiotics. Both have same antibacterial spectrum and similar mechanism of action as erythromycin.

NEWER MACROLIDE ANTIBIOTICS

In an attempt to overcome the limitations of erythromycin and other older macrolide antibiotics like gastric acid labiality, short half-life, and narrow antimicrobial spectrum, a number of semi-synthetic macrolides have been developed. These include azithromycin, clarithromycin, and roxithromycin.

Macrolides are effectively used for Bacterial Kidney Disease (BKD) caused by *Renibacterium* sp.

LINCOSAMIDES

Lincosamides are a group of manoglycoside antibiotics containing an amino acid like side Chain. They resemble macrolide antibiotics in several aspects and are often considered with macrolides. Lincosamides have limited use in human and

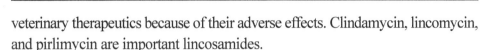

veterinary therapeutics because of their adverse effects. Clindamycin, lincomycin, and pirlimycin are important lincosamides.

Chemistry and properties

Lincosamides are monobasic compounds that are considered derivatives of an amino acid and a Sulphur containing octose. Like macrolides, they are more stable in salt forms. They are basic compounds, so are more concentrated in acidic fluids.

Mechanism of action

Lincosomides inhibit the bacterial protein synthesis and are primarily bacteriostatic. They bind exclusively to the 50 S ribosomal subunit in a manner similar to the macrolides. Although lincosamides, macrolides, and chloramphenicol are not structurally related, they all act at sites with close proximity. The overlapping of binding sites leads to competition for the binding site, thus reducing the efficacy of these drugs on concurrent use. Similar to macrolides, lincosamides are more active at an alkaline pH and may become bactericidal at high concentration.

Antibacterial spectrum

Lincosamides have antibacterial spectrum similar to that of macrolides. In general, lincosamides are more active against Gram-positive organisms with anaerobes showing more susceptibility than aerobes. Gram-negative bacteria tend to be resistant to lincosamides. Clindamycin has a broader spectrum than lincomycin and is effective against some Gram-negative anaerobes. Lincosamides are also active against Toxoplasma and some strains of Mycoplasma.

Pharmacokinetics

Lincosamides are absorbed to variable extent (60-90%) after oral administration. Absorption from IM injection site is good and rapid. After absorption, lincosamides are widely distributed in body fluids and tissues with appreciable concentrations in bones and soft tissues. Metabolic inactivation of lincosamides is variable. Some of the metabolites retain antibacterial activity and may prolong duration of action. The parent drug and its active and inactive metabolites are excreted in bile and urine; the extent of excretion depends on the type of compound and route of administration.

Lincomycin

Lincomycin is a naturally occurring lincosamide antibiotic obtained from *Streptomyces lincolensis*. It occurs as white to crystalline powder that is freely soluble in water. It is more stable in salt form and is commercially available as monohydrate hydrochloride.

Antibacterial spectrum

Lincomycin is primarily effective against Gram-positive bacteria, anaerobes, and Mycoplasma, but has little activity against Gram-negative organisms. It is primarily bacteriostatic but tends to become bactericidal at high concentrations against highly susceptible organisms.

Pharmacokinetics

Lincomycin is absorbed to the extent of 60% from GI tract following oral dosing. Peak plasma concentrations of lincomycin after oral administration occurs at 2-4 hours after oral dosing. Absorption from IM injection site is rapid with peak plasma levels occurring at 1-2 hours. It is widely distributed in body tissues and fluids. Lincomycin is partly (~50%) metabolized in liver and excreted as unchanged drug and metabolites in bile and urine. The elimination half-life of lincomycin in small animals is about 3-4 hours.

Clindamycin

Clindamycin is a semi-synthetic derivative of lincomycin (7-chloro-7-deoxy lincomycin). Chemical substitution of a chlorine atom for a-hydroxyl group allows clindamycin to be more potent, better absorbed and less toxic than the parent drug lincomycin. In general, clindamycin and lincomycin have same antibacterial spectrum, but clindamycin is effective against sensitive bacteria in considerably lower concentrations and is cidal for more strains of bacteria.

Pirlimycin

Pirlimycin is a lincosamide antibiotic with activity similar to other members of the family. It is useful against infections caused by Gram-positive bacteria *Staphylococcus aureus* and *Streptococcus* spp.

10

MISCELLANEOUS ANTIBACTERIALS

A number of antibacterial drugs are used in therapeutics but they do not fit properly into important categories of drugs, some of these antibacterial are not commonly used or available. These miscellaneous antibacterial drugs are discussed in the present chapter.

Glycopeptide Antibiotics

Glycopeptides are glycosylated cyclic or polycyclic peptides produced by various group of filamentous actinomycetes. These therapeutics target Gram-positive bacteria by binding to the acyl-D-Ala-D-Ala terminus to the growing peptidoglycan and then cross-linking peptides within and between peptidoglycan on the outer surface of the cytoplasmic membrane. Glycopeptide antibiotics are often used to treat life threatening infections by multi-drug-resistant Gram-positive organisms, such as *S. aureus*, *Enterococcus* spp., and *Clostridium difficile.* Natural glycopeptides are composed of a cyclic peptide core with seven amino acids, to which two amino sugars are bound to the amino acid core.

Glycopeptide antibiotics include

- Vancomycin
- Teicoplanin

- Telavanin

- Dalbavancin

Vancomycin, produced by the actinomycete *Amycolatopsis orientalis*, was first introduced into clinical practice in 1958. Vancomycin is a drug of last resort, used only after treatment with other antibiotics had failed. It works by preventing the incorporation of NAM-NAG peptide subunits from being incorporated into the peptidoglycan matrix.

Teicoplanin is a glycopeptide antibiotic that is produced by the actinomycete *Actinoplanes teichomyceticus*. Teicoplanin and vancomycin have a similar mode of many chemical and microbiological properties, but teicoplanin has longer elimination half-life and the possibility of administration by intramuscular injection. Teicoplanin is a complex of six analogous molecules. As in vancomycin, the core aglycone is a cyclic heptapeptide backbone consisting of aromatic amino acid residues and carries two sugar moieties D-mannose and N-acetyl-β-Dglucosamine and a fatty-acid chain. The fatty-acid component increases teicoplanin's lipophilicity, resulting in greater cellular and tissue penetration.

Glycopeptide antibiotics inhibit synthesis of the bacterial cell wall by binding to the dipeptide terminus D-Ala-D-Ala of peptidoglycan precursors, thereby sequestering the substrate from transpeptidation reactions at the late extracellular stages of peptidoglycan crosslinking. The complex of D-Ala-D-Ala with glycopeptides is stabilized by an arrangement of hydrophobic van der Waals bonds and five hydrogen bonds lining the antibiotic-binding pocket. Cross-linked peptidoglycans are needed for sufficient tensile strength of the cell wall. Thus, a glycopeptide's action finally destabilizes the cell wall, and the bacterial cell death occurs presumably due to osmotic damage.

The spectrum of teicoplanin's activity against Gram-positive bacteria is similar to that of vancomycin, but teicoplanin has greater potency, particularly against some clinical microbe of the genera Staphylococcus, Streptococcus, and Enterococcus.

Semisynthetic glycopeptides Telavancin and Dalbavancin have been developed to overcome the emergence of MRSA strains showing weak sensitivity to vancomycin and to increase the penetration into tissues. Consequently, most of the semisynthetic glycopeptides were created by introducing hydrophobic moieties into the heptapeptide scaffold to ensure the membrane-anchoring ability, thus leading to more effective drugs. Telavancin is a derivative of vancomycin and differs in that the majority of the molecules are associated with the cell membrane integrity rather than the cell wall biosynthesis. Dalbavancin is a semi synthetic antibiotic derived from a Teicoplanin analog *via* modification of the functional groups and sugar moieties, without disruption of the D-Ala-D-Ala-binding site, which is required for antimicrobial activity. Dalbavancin prevents the synthesis of the bacterial cell wall by binding to the D-Ala-D-Ala residues of growing peptidoglycan chains and thus inhibits/disrupts peptidoglycan elongation and cell membrane formation. Compared to vancomycin, dalbavancin shows a very potent *in vitro* activity against the majority of Gram-positive pathogenic bacteria

Polypeptide antibiotics

Polypeptide antibiotics include bacitracin, polymyxin B and polymyxin E (colistin). Bacitracin is a polypeptide antibiotic that inhibits cell wall synthesis and is active against Gram-positive bacteria. Colistin (polymyxin E) and polymyxin B are cationic polypeptide antibiotics that disrupt the outer bacterial cell membrane by binding to the anionic outer membrane and causing bacterial cell death.

Bacitracin

The lipid carrier involved in transporting the cell wall building block across the membrane is bactophenol (C55 isoprenyl phosphate). The lipid acquires an additional phosphate group in the transport process and must be dephosphorylated in order to regenerate the native compound for another round of transfer. The cyclic peptide antibiotic, bacitracin binds to the C55 lipid carrier. Bacitracin inhibits its dephosphorylation, thereby inhibiting cell wall synthesis.

Polymyxin

The two polymyxins that are used clinically are polymyxin B and polymyxin E (also known as colistin). These compounds, which were originally isolated in 1947 from a *Bacillus polymyxa*. Polymyxin B and colistin both are rapidly bactericidal against susceptible organisms, binding to the lipopolysaccharides and phospholipids in the outer cell membrane of Gram-negative bacteria. This displaces calcium and magnesium from the phosphate group of membrane lipids, which leads to disruption of the outer cell membrane, resulting in leakage of intracellular contents, and finally bacterial death.

Nitofurans

Nitofurans are synthetic antimicrobial agents with a broad spectrum activity covering not only Gram-positive and gram-negative but also several types of protozoan parasites. They are normally bacteriostatic but can be bactericidal at high doses. Most nitrofurans are very poorly absorbed from gatro-intestinal tract. Some nitrofurans not absorbed through gastro-intetinal tract are absorbed from water and can be administered by immersion. Two nitrofurans most commonly used in fish medicine are Furazolidone and Nifurpirinol.

11

ANTIFUNGAL AGENTS

Fungal infections are caused by microscopic organisms that can invade the epithelial tissue. The fungal kingdom includes yeasts, molds, rusts and mushrooms. Fungi, like animals, are heterotrophic, that is, they obtain nutrients from the environment, not from endogenous sources (like plants with photosynthesis). Most fungi are beneficial and are involved in biodegradation, however, a few can cause opportunistic infections if they are introduced into the skin through wounds, or into the lungs and nasal passages if inhaled. Diseases caused by fungi include superficial infections of the skin by dermatophytes and systemic infections (deep infections) of the internal organs. The dermotophytic infections are named after the site of infection rather than the causative organism.

In the last 50 years, there was a steady increase in fungal infections attributed mainly to intensive use of broad-spectrum antibacterials. Depending on the requirement, initially two important antibiotics – Amphotericin B and Griseofulvin were introduced around 1960. This was followed by Flucytosine and Imidazoles in 1970s and Triazoles in 1980s. Some newer compounds like terbinafine have been introduced recently.

Fungi cell wall

Fungi are eukaryotic organisms, have membrane bound organelle. They have a cell membrane surrounded by a rigid cell wall. Fungal cell membrane is similar to that of other eukaryotes, lipid bilayer with protein embedded within it. A major component of the eukaryotic cell membrane is sterols, which are virtually absent in all prokaryotes such as bacteria. Ergosterol is the essential sterol in fungal cell membrane. 60-70% of fungal cell wall is made up of polysaccharides. Glucan is the major polysaccharide present in fungal cell wall. Cell wall of fungi have higher content of chitin.

Biochemical Targets for Antifungal Chemotherapy

Fungal cells are complex organisms that share many biochemical targets with other eukaryotic cells. Therefore, agents that interact with fungal targets not found in eukaryotic cells are needed. The fungal cell wall is a unique organelle that fulfils the criteria for selective toxicity. The fungal cell wall differs greatly from the bacterial cell wall. Fungi being eukaryotes, very similar to eukaryotic cells, it is very hard to find the drug which are selectively toxic. One of the main target for the antifungal therapy is the microbial plasma membrane. Eukaryotic cell membranes are pretty similar to each other, have sterols (they are special lipids that can regulate membrane fluidity). Cholesterol is the main sterol in animal cell membrane. Fungal cells have ergosterol instead. So the antifungal drugs make use of this.

Arrangement of the biomolecular components of the cell wall accounts for the individual identity of the organism. Although, each organism has a different biochemical composition, their gross cell wall structure is similar. There are three general mechanisms of action for the antifungal agents: cell membrane disruption, inhibition of cell division and inhibition of cell wall formation.

Classification of Antifungal Drugs

I.   Antifungal antibiotics

    1.   Polyenes: eg., Amphotericin B, Nystatin

    2.   Heterocyclic benzofurans: eg., Griseofulvin

II. Antimetabolites: eg., Flucytosine

III. Azoles

  1. Imidazoles: eg., Ketaconazole, Miconazole

  2. Triazoles: eg., Fluconazole, Itraconazole

IV. Allylamines: eg., Terbinafine

V. Echinocandins: eg., Caspofungin

I. Polyene antibiotics

Polyene antibiotics contain a large ring of atoms (essentially a cyclic ester ring) with multiple conjugated carbon-carbon double bonds (hence polyene) on one side of the ring and multiple hydroxyl groups bonded to the other side of the ring. These polyene antimycotics are obtained from *Actinomycetes*. The polyenes bind to ergosterol in the fungal cell membrane contribute to fungal cell death. They are poorly soluble in water and common organic solvents, but are soluble in highly polar solvents such as DMSO.

Amphotericin B

Amphotericin B (AMB) was the first antifungal drug developed and is approved for the treatment of many invasive fungal infections. It is an amphoteric polyene antibiotic obtained from *Streptomyces nodustus*. Two amphotericins, amphotericin A and amphotericin B are known, but only B is used clinically because it is significantly more active *in vivo*. Amphotericin A is almost identical to amphotericin B, but has little antifungal activity. Amphotericin B is insoluble in water and its solution is unstable and rapidly decomposes on exposure to light.

Chemistry

It is a polyene macrolide antibiotic, more precisely a heptane macrolide containing seven conjugated double bonds in the transposition and 3-amino-3,6-dideoxymannose connected to main ring by a glycosidic bond. One side of the macrocyclic ring that contains conjugated double bonds is highly lipophilic, while the other side containing many –OH groups is hydrophilic. This structure provides amphoteric nature to the compound, for which the drug is named.

Mechanism of action

The amphotericin B and other polyenes have affinity for sterols, particularly ergosterol, present in fungal cell membranes. The interaction of AMB and other polyene antibiotics with membrane ergosterol results in formation of channels or pores in the cell membrane with altered permeability and leakage of cellular contents. The altered K+/H+ exchange results in efflux of potassium and influx of hydrogen ions, producing a state of acidosis that halts some important enzymatic processes. Loss of important organic molecules such as amino acids and sugars from the fungal cell results in irreversible damage. High concentration of amphotericin B directly disrupts the fungal cell membrane permeability.

Antimicrobial spectrum

Amphotericin β and other polyene antibiotics have broad spectrum antifungal activity, although sensitivity of various species and strains of fungi to these antibiotics vary widely. It is useful against various systemic fungi including *Candida, Cryptococcus, Aspergillus.* Some alage and protozoa (*Trypanosoma, Tricomonas*) are sensitive to polyene antibiotics.

Pharmacokinetics

Amphotericin B is poorly absorbed from GI tract and therefore, oral AMB is used only for gastrointestinal fungal infection. Amphotericin B is extensively bound to plasma lipoproteins. It appears to become associated with cholesterol containing membranes in many tissues from which it is slowly released into the circulation. The elimination of Amphotericin B is complicated and not clearly defined due to binding of the drug to tissues and cholesterol.

Adverse effects

Amphotericin B is toxic drug with side effects, especially renal function impairment. The nephrotoxicity occurs *via* binding of drug to membrane cholesterol in the renal tubular cell membrane. Interaction of drug with renal membrane cholesterols produce alterations in tubular membrane permeability leading to altered electrolyte fluxes.

Drug interactions

Amphotericin B may be combined with other antimicrobial agents such as 5-flucytosine to produce synergistic effect. Amphotericin B should not be combined with aminoglycosides due to nephrotoxic effects.

Nystatin

Nystatin is a polyene antifungal produced by *Streptomyces noursei*. It is structurally similar to Amphotericin B and has the same mechanism of action.

2. Heterocyclic benzofurans

Griseofulvin

It is a systemic antifungal obtained from *Penicillium griseofulvin*. It is an odourless, bitter tasting, white to creamy powder that is very slightly soluble in water and slightly soluble in most organic solvents.

Mechanism of action

Griseofulvin is a fungistatic drug that enters into the susceptible fungi through energy dependent transport system. It then acts by interfering with the polymerisation of the microtubular protein with microtubules. Interaction with microtubules interferes with the spindle formation in dividing cells thereby arresting the metaphase of cell division. This leads to production of multinucleate fungal cells. Impairment of microtubule function may also interfere with the transport of essential material through the cytoplasm to periphery, which accounts for inhibition of hyphal cell wall synthesis.

Antimicrobial spectrum

Griseofulvin is a narrow-spectrum antifungal agent active only against dermatophytes. It is ineffective against deep mycoses. It is fungistatic against older and dormant fungi, but may kill (fungicidal) actively metabolising and growing young cells.

Pharmacokinetics

After oral administration, Griseofulvin is variably and erratically absorbed from GI tract mainly due to its poor soubility. After absorption, Griseofulvin is concentrated in skin, fat, skeletal muscle and liver and metabolised by the liver to inactive metabolites. About one-half of the drug is excreted as metabolites in urine and the remaining is excreted unchanged in faeces.

II. Antimetabolites

Flucytosine mechanism of action

Flucytosine (5- Flurocytosine) is afluorinated pyrimidine antifungal agent. Flucytosine, an analogue of cytosine, is inactive as such and requires conversion into active metabolites inside the fungal cells. On administration, Flucytosine enters fungal cells *via* a cytosine-specific permease, an enzyme not found in animal cells. Inside the fungal cell, it is rapidly converted into 5-Fluorouracil by cytosine deaminase enzyme. The 5-fluoruracil acts as an antimetabolite by competing with uracil. It initially forms 5-fluorouracil monophosphate (5-fluorouracil ribose monophosphate) and then 5- fluorouracil triphosphate, which interferes with pyrimidine metabolism and eventually RNA and protein synthesis. 5-fluorouracil monophosphate by an alternate pathway is also metabolized to 5-fluorodeoxyuracil monophosphate, an inhibitor of thymidylate synthetase. Inhibition of thymidilate synthetase deprives the organisms of thymidylic acid, an essential DNA component, which eventually disrupts DNA synthesis and cell division.

The combination of flucytosine and amphotericin B is synergistic because amphotericin affects fungal cell permeability, allowing more of flucytosine to penetrate the cell.

Fungal resistance

Fungal resistance to flucytosine develops rapidly even during course of treatment. The rate of emergence of resistance is lower with a combination of amphotericin B and flucytosine that it is with flucytosine alone. Although the mechanism of resistance is not exactly known, decreased levels of any of the enzymes required

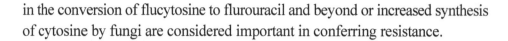

in the conversion of flucytosine to flurouracil and beyond or increased synthesis of cytosine by fungi are considered important in conferring resistance.

Pharmacokinetics

Flucytosine is rapidly and almost completely absorbed after oral administration. It is distributed widely throughout the body. Only about 2-4% of the drugs is bound to plasma protein. About 80-95% of the drug is excreted *via* urine by glomerular filtration.

III. Azoles

Azole antifungal agents are the largest class of synthetic antimycotics. About 20 agents are on the market today. Some are used topically to treat superficial dermatophytic and yeast infections. Others are used systemically to treat severe fungal infections. Azoles are group of five membered heterocyclic compounds containing one or more nitrogen atoms in the ring; the number of nitrogen atoms present being indicated by a prefix, as in triazole. These compounds possess significant broad spectrum antifungal actions with similar mechanism of action and antifungal spectrum. The important antifungal azole derivatives further consist of two subgroups – Imidazoles and Triazoles.

1. Imidazoles

Imidazoles are a group of organic compounds in which the aromatic heterocyclic in a diazole. They are active against many microorganisms and parasites including fungi, bacteria, helminths and protozoa. They are generally poorly soluble in water but can be dissolved in organic solvents.

Mechanism of action

The Imidazoles and Triazoles act on the fungal cell membrane and alter the membrane permeability of susceptible fungi by inhibition of ergosterol synthesis. They inhibit 14α-demethylase, a fungal microsomal cytochrome 450 dependent enzyme. This enzyme normally participates in the sterol biosynthesis pathway and catalyses demethylation of lanosterol to ergosterol, the major cell membrane component. The reduced synthesis of ergosterol content in the cell membrane in response to azole antifungals in turn decreases fluidity of membrane and increases the permeability with effects similar to Amphotericin B. Inhibition of ergosterol synthesis also results in accumulation of 14α-methyl sterols, which impair membrane functions leading to alterations in energy metabolism and growth inhibition. Azoles are also reported to affect activation of oxidative and peroxidative enzymes resulting in intracellular accumulation of toxic levels of hydrogen peroxide. The overall effect of imidazoles on fungal and other microbial cells is cell membrane and internal organelle disruption and cell death. Cholesterol synthesis in mammals is not affected as it does not require 14α-demethylase.

2. Triazoles

The triazoles are broad spectrum azole antifungal agents which are usually considered along with imidazoles because both groups share same mechanism of action and spectrum. However, triazole antifungal agents are more slowly metabolized and have less effect on animal sterol synthesis than do the imidazoles.

IV. Allylamines

Terbinafine

It is a synthetic drug that belongs to a new allylamine class of antifungal agents. It is a white fine crystalline powder that is freely soluble in methanol, soluble in ethanol and slightly soluble in water. It is highly lipophilic in nature and tends to accumulate in skin and fatty tissues.

Mechanism of action

Terbinafine is a fungicidal drug that acts by selectively inhibiting squalene epoxide enzyme, which is involved in the synthesis of ergosterol from squalene in the fungal cell wall. Thus, terbinafine decreases synthesis of ergosterol in cell wall and causes accumulation of squalene within cell of susceptible fungi. Fungal cell death usually results from disruption of cell membrane and accumulation of toxic squalene.

Pharmacokinetics

It is well absorbed after oral administration, but it undergoes significant first pass metabolism that decreases bioavailability to about 40%. It is highly plasma protein bound (99%). Its lipophilic nature permits wide distribution in tissues. It is metabolised in the liver by hepatic cytochrome P-450. The metabolites are excreted in urine (80%) and feces (20%).

V. Echinocandins

Echinocandins are a newer group of antifungal drugs which inhibit synthesis of fungal cell wall and are thus called penicillins of antifungal drugs. Important agents of the group include caspofungin.

Caspofungin

Caspofungin is a water soluble, semi-synthetic lipopeptide antifungal drug synthesized from fermentation product of *Glarea lozoyensis*.

Mechanism of action

Caspofungin has a novel mechanism of antifungal action. It inhibits the synthesis of β-glucan in the fungal cell wall, probably *via* non-competitive inhibition of the enzyme, 1,3 β-glucan synthase. B-glucans are polymers which when linked in their tens or thousands, make up cell wall membranes. Inhibition of glucan by echinocandins hence weakens and disturbs the integrity of the fungal cell wall. Caspofungin and other echinocandins are relatively selective in their action because glucan synthase enzyme is not present in animal cells.

12

ANTIPARASITIC DRUGS

Antiparasitics are a class of medications which are indicated for the treatment of parasitic diseases, such as those caused by helminths, ectoparasites, and protozoa. Antiparasitics target the parasitic agents of the infections by destroying them or inhibiting their growth, they are usually effective against a limited number of parasites within a particular class. Any organism that lives in or on another larger organism of a different species upon which it depends for food is called parasite. Although parasite benefits from association, the host is harmed but they generally don't kill their host. There are two major groups of parasites: multicellular helminths (worms) and single-celled protozoa.

Characteristics of an ideal anti-parasitic drug

1. A wide therapeutic index
2. Broad spectrum of activity against mature & immature larval worms of most types of parasites
3. Effective orally
4. Effective in a single dose
5. Leave no or low tissue residues
6. Economic (inexpensive) and compatible with other drugs.
7. Inhibit re-infection for extended periods.

Antiprotozoal Drugs

Antiprotozoal drug is any agent that kills or inhibits the growth of protozoans. Protozoans typically are microscopic, they are similar to plants and animals in that they are eukaryotes and thus have a clearly defined cell nucleus. This distinguishes them from prokaryotes, such as bacteria. As a result, many of the antibiotics that are effective in inhibiting bacteria are not active against protozoans.

Fumagillin

Fumagillin is an antibiotic produced by the parasitic fungus, *Aspergillus fumigatus,* and acts by inhibiting RNA synthesis. It is acidic and is normally presented as dicyclohexylamine (DCH) salt. This salt is sparingly soluble in water and soluble in ethanol. Fumagillin DCH is heat-liable so it cannot be incorporated in feed before pelleting. The pure drug may be surface-coated onto pellets by dissolving in 95% ethanol and spraying the solution onto the pellets. A 2% premix (Fumidil B®) is available commercially, and this may be surface coated onto pellets by conventional methods.

It has low antibacterial and antifungal activity and is primarily active against protozoa. Fumagillin DCH has been used for the treatment of diseases *viz.,* *Enterocytozoon salmonis* in Chinook salmon, *Pleistophora anguillarium* in eels, *Sphaerospora renicola* in common carp, *Myxosoma cerebralis* in rainbow trout.

Nitroimidazoles

The nitroimidazoles are a group of synthetic antimicrobial agents active against protozoa and obligate anaerobic bacteria. In this, dimetridazole is used for the treatment of 'white spot' (*Ichthyophthirius multifilis* infection or 'ich') in rainbow trout. Nitroimidazoles have been recommended for bath treatment of some ornamental species, especially for 'hole-in-the-head'disease, initiated by the protozoan, *Hexamita.* Where malachite green is not available, nitroimidazoles may also be used for 'velvet disease' caused by protozoan *Oodinium.*

Nitroimidazoles are active on protozoa whose energy metabolism is anaerobic. These organelles produce ATP from the conversion of pyruvate to acetate, CO_2 and H_2. Reduction of nitroimidazole's by ferredoxin produces nitro radicals which then attack the parasite's DNA. Interestingly, nitroimidazoles are also active and clinically used against anaerobic bacteria. Mammalian cells don't produce large amounts of nitroimidazole radicals, and in addition are able to detoxify them through reaction with molecular oxygen. The superoxide formed in the latter reaction is then scavenged by superoxide dismutase.

It is also recommended for bath treatment of some ornamental species, especially for 'hole-in-the-head' disease by the protozoan, Hexamita by dipping in 7 ppm metronidazole. Nitroimidazole may also be used for 'velvet disease' caused by the protozoan, *Oodinium* sp., but twice the concentration used for hole in the head disease is recommended.

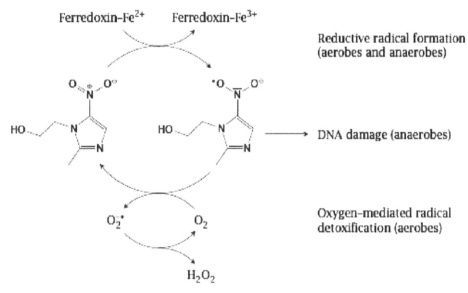

Anthelmintic Drug

Infection by *helminths* (worms) may be limited solely to the intestinal lumen or may involve a complex process with migration of the adult or immature worm through the body before localization in a particular tissue. Complicating our understanding of the host– parasite relationship and the role of chemotherapy in helminth-induced infections is the complex life cycle of many of these organisms. Whereas some helminths have a simple cycle of egg deposition and development of the egg to produce a mature worm, others must progress through one or more hosts and one or more morphological stages, each metabolically distinct from the other, before emerging as an adult. Furthermore, an infective form may be either an adult worm or an immature worm. Treatment may be further complicated by infection with more than one genus of helminth.

Pathogenic helminths can be divided into the following major groups: *Cestodes* (flatworms), *Nematodes* (round-worms), *trematodes* (flukes) and less frequently, *Acanthocephala* (thorny-headed worms).

Properties of an ideal antihelmitic

Most available anthelmintic drugs exert their antiparasitic effects by interference with (1) energy metabolism, (2) neuromuscular coordination, (3) microtubular function, and (4) cellular permeability.

Treatment for infections caused by nematodes

Nematodes are long, cylindrical unsegmented worms that are tapered at both ends. Because of their shape, they are commonly referred to as roundworms. Some intestinal nematodes contain a mouth with three lips, and in some the mouth contains cutting plates. Infection occurs after ingestion of embryonated eggs or tissues of another host that contain larval forms of the nematodes.

Benzimidazoles

Several benzimidazoles are in use for the treatment of helminthic infections. Three of them, Mebendazole, Thiabendazole and Albendazole are important in treating helmintic infections. They have a broad range of activity against many nematode and cestode parasites.

Thiabendazole

Thiabendazole (*Mintezol*) inhibits fumarate reductase and electron transport–associated phosphorylation in helminths. Interference with ATP generation decreases glucose uptake and affects the energy available for metabolism. Benzimidazole anthelminthics (e.g., thiabendazole, mebendazole and albendazole), bind selectively to tubulin of nematodes (roundworms), cestodes (tapeworms), and trematodes (flukes). This inhibits microtubule assembly, which is important in a number of helminth cellular processes, such as mitosis, transport and motility.

Mebendazole

Unlike Thiabendazole, Mebendazole (*Vermox*) does not inhibit fumarate reductase. While mebendazole binds to both mammalian and nematode tubulin, it exhibits a differential affinity for the latter, possibly explaining the selective action of the drug. The selective binding to nematode tubulin may inhibit glucose absorption, leading to glycogen consumption and ATP depletion.

Albendazole

Albendazole appears to cause cytoplasmic microtubular degeneration, which in turn impairs vital cellular processes and leads to parasite death. There is some evidence that the drug also inhibits helminth-specific ATP generation by fumarate reductase.

Piperazine

Piperazine (*Vermizine*) contains a heterocyclic ring that lacks a carboxyl group. It works by causing a paralysis within the muscles lining the body wall of these worms. Piperazine acts as an agonist at gated chloride channels on the parasite muscle. It acts on the musculature of the helminths to cause reversible flaccid paralysis mediated by chloride-dependent hyperpolarization of the muscle membrane. This is achieved by mimicking the neurotransmitter to induce the relaxation of these muscles which results in expulsion of the worm.

Uses

Piperazine is an organic compound used as an anti-parasitic in veterinary medicine, primarily for worms. Piperazine works through anthelmintic action (used to expel or destroy parasitic worms in the gastro-intestinal tract). It should never be used in the presence of invertebrates as for the reasons outlined above. Piperazines aquatic uses are restricted to internal parasite control, especially intestinal worms. Piperazine is proven effective for *Capillaria* Nematode worms that infest the intestines of Angelfish, Discus, some other cichlids, and occasionally other aquarium species. In the aquarium the disease spreads easily from fish to fish as they consume the eggs of the worms, shed in feces of infected individuals.

Ivermectin

Ivermectin (*Mectizan*) acts on parasite-specific inhibitory glutamate-gated chloride channels that are phylogenetically related to vertebrate GABA-gated chloride channels. Ivermectin causes hyperpolarization of the parasite cell membrane and muscle paralysis. At higher doses it can potentiate GABA-gated chloride channels. It does not cross the bloodbrain barrier and therefore has no paralytic action in mammals. Ivermectin is administered by the oral and subcutaneous routes. It is rapidly absorbed and most of the drug is excreted unaltered in the feces. The half-life is approximately 12 hours. 0.2 mg/kg dose is effective when given in feed.

Pyrantel Pamoate

Pyrantel pamoate (*Antiminth*) is a agonist at the nicotinic acetylcholine receptor, and its actions result in de-polarization and spastic paralysis of the helminth muscle. Its selective toxicity occurs primarily because the neuromuscular junction of helminth muscle is more sensitive to the drug than is mammalian muscle. This drug is administered orally, and because very little is absorbed, high levels are achieved in the intestinal tract. Less than 15% of the drug and its metabolites are excreted in urine.

Levamisole

Levamisole as an antiparasitic agent appears to be tied to its agnositic activity towards the nicotinic acetylcholine receptors in nematode muscles. This agonistic action reduces the capacity of the males to control their reproductive muscles and limits their ability to copulate. The effects of levamisole on the immune system are complex. The drug appears to restore depressed immune function rather than to stimulate response to above-normal levels. Levamisole can stimulate formation of antibodies to various antigens, enhance T-cell responses by stimulating T-cell activation and proliferation, potentiate monocyte and macrophage functions including phagocytosis and chemotaxis, and increase neutrophil mobility, adherence, and chemotaxis.

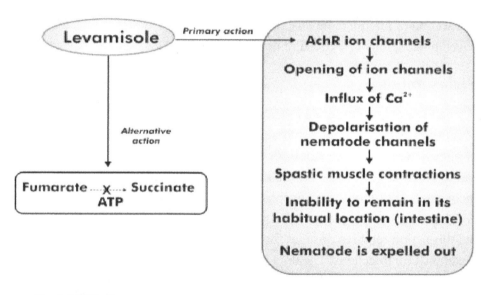

X - Inhibitation
········► Convertion

Dosage: If Levamisol cannot be located, the use of Levamisol HCl found in many commercial dog, poultry, cattle, etc. drugs can be substituted. 59 mg of Levamisole HCl is equivalent to 50 mg of pure levamisole.

About 2.36 mg/L (or 9 mg/Gallon) of Levamisole HCl is needed. So, approximately 90mg of Levamisole HCl is needed for 10 gallons (38 liters) to make the required concentration of 2 ppm.

Treatment for infections caused by cestodes

Cestodes, or tapeworms, are flattened dorsoventrally and are segmented. The tapeworm has a head with round suckers or sucking grooves. Some tapeworms have a projection (*rostellum*) that bears hooks. This head, or *scolex* (also referred to as the hold-fast organ), is used by the worm to attach to tissues. Drugs that affect the scolex permit expulsion of the organisms from the intestine. Attached to the head is the neck region, which is the region of growth. The rest of the worm consists of a number of segments, called *proglottids,* each of which contains both male and female reproductive units. These segments, after filling with fertilized eggs, are released from the worm and discharged into the environment. Cestodes have complex life cycles, usually requiring development in a second or intermediate host. Following their ingestion, the infected larvae develop into adults in the small intestine.

Niclosamide

For many years, niclosamide (*Niclocide*) was widely used to treat infestations of cestodes. Niclosamide is a chlorinated salicylamide that inhibits the production of energy derived from anaerobic metabolism. It may also have adenosine triphosphatase (ATPase) stimulating properties. Inhibition of anaerobic incorporation of inorganic phosphate into ATP is detrimental to the parasite. Niclosamide can uncouple oxidative phosphorylation in mammalian mitochondria, but this action requires dosages that are higher than those commonly used in treating worm infections. The drug affects the scolex and proximal segments of the cestodes, resulting in detachment of the scolex from the intestinal wall and eventual evacuation of the cestodes from the intestine by the normal peristaltic action of the host's bowel. Because niclosamide is not absorbed from the intestinal tract, high concentrations can be achieved in the intestinal lumen. Niclosamide has been used extensively in the treatment of tapeworm infections caused by *Diphyllobothrium latum.*

Treatment for infections caused by trematodes

Trematodes (flukes) are non-segmented flattened helminths that are often leaf like in shape. Most have two suckers, one found around the mouth (oral sucker) and the other on the ventral surface. Most are hermaphroditic. The eggs, which are passed out of the host in sputum, urine, or feces, undergo several stages of maturation in other hosts before the larvae enter host. The larvae are acquired either through ingestion of food (aquatic vegetation) or by direct penetration of the skin. After ingestion, most trematodes mature in the intestinal tract (intestinal flukes); others migrate and mature in the liver and bile duct (liver flukes), whereas still others penetrate the intestinal wall and migrate through the abdominal cavity to the lung (lung flukes).

The *schistosomes* (blood flukes) are a distinct group of trematodes. These helminths are cylindrical at the anterior end and flattened at the posterior end. The sexes are separate. The larvae penetrate skin that is in contact with contaminated water and then migrate through the lymphatics and blood vessels to the liver. After maturing, schistosomes migrate into the mesenteric or vesicular vein, where the adults mate and release eggs. The eggs secrete enzymes that enable them to pass through the wall of the intestine.

Praziquantel

The neuromuscular effects of praziquantel (*Biltricide*) appear to increase parasite motility leading to spastic and paralysis. The drug increases calcium permeability through parasite specific ion channels, so tegumental and muscle cells of the parasite accumulate calcium. Insertion of the drug into fluke's lipid bilayer causes conformational changes, rendering the fluke susceptible to antibody or complement mediated assault. Praziquantel is an extremely active broad-spectrum anthelminthic. It is the most effective of the drugs used in the treatment of schistosomiasis. Mixing with isoniazid, pyrazinamide, or rifampin significantly decrease the effective blood levels of praziquantel rendering it ineffective.

Dyes

Dye chemicals from the family of the triphenylmethane dyes, malachite green (MG), a common commercial and inexpensive fabric dye, has developed and been used as a therapeutic multi-usage drug to globally reduce parasitic, microbial, and fungal diseases found in fish and seafood farming. MG has been used both prophylactically and in the treatment of fungal infections for fish and eggs for more than 80 years. In the course of the 1980s, 1990s, and 2000s, many concerns were raised in regard to the toxicity of this substance, and different toxicological studies were carried out for MG and for some other similar dyes applied or potentially applied for their therapeutic qualities in fish farming. MG has now been banned in nearly all of the regions of the world, including North America and Europe, but can still be present in various inappropriate fish farming practices around the world. In India MG is being used as a prophylactic in ornamental fish farms.

Recently, the Joint WHO/FAO Expert Committee on Food Additives (JECFA) has evaluated the risk for public health of the use of MG and crystal (gentian) violet (CV) in fish farming. The Codex Committee on Residues of Veterinary Drugs in Foods has recommended that competent authorities should not permit their use in food-producing animals including fish/seafood farming. This should therefore lead to an absence of detectable residues in products from this industry. However, they still appear to be present, probably because they are still widely used in the textile industry and elsewhere and are commercially available as inexpensive therapeutic chemicals for ornamental fish. In addition, the dyes are persistent in the sediment of water sources for aquaculture and will be absorbed and bioaccumulated in fish tissues over time.

Dyes with pharmacological activity can be categorized into five chemical classes: triaryl(phenyl)methanes, phenothiazines, xanthenes, acridines, and azo compounds. In aquaculture, dyes are primarily used as a treatment for fungal and external parasite infections in fish and to protect incubating eggs from fungus. Many of the dyes described from these chemical classes have antiseptic, antimicrobial, or other medicinal properties with uses in veterinary and human medicine.

Triarylmethanes

Malachite Green, Crystal Violet and Brilliant Green are in the triarylmethane dye class of dyes. Triarylmethane dyes are cationic and have wide application as colorants for textiles, papers, plastics, and inks and are used as biological stains. These are characterized as the structurally simple triphenylmethane dyes. The triphenylmethane dyes have a long history of therapeutic use as fungicide and ectoparasiticide agents. MG to be unusually effective for the treatment of fungus infections in trout, bass, and trout eggs. MG is considered to be the most effective antifungal treatment used in aquaculture. Exposure bath treatments are effective for the control of the external protozoan *Ichthyophthirius multifiliis* in fish, and treatments of fish eggs with dilute MG effectively reduce fungal growth (e.g., *Saprolegnia*) and ensure viability of live eggs. CV and BG are other triphenylmethane dyes with similar properties to MG.

Malachite green (Also known as Analine Green, Victoria Green)

A bacteriological stain and topical antiseptic used to treat parasites, and fungal infections in fish and fish eggs. The chemical formula is $C_{23}H_{25}ClN_2$. Malachite Green is often combined with formaldehyde (aka formalin) and is somewhat similar to methylene blue in terms of uses. However, it is a harsher chemical and must be used with caution. It is the most effective against external parasites, particularly when combined with formaldehyde. Disorders treated by Malachite Green include the following:

- *Egg Fungus* - Serves as a safe disinfectant for fish eggs (0.05 ppm).
- *Ichtyophthirius multifilis* - White spot or Ich of fish
- Velvet disease caused by the parasite, Oodinium .
- Saprolegnia, a freshwater mold that attacks fish eggs.
- Do not use full strength with sensitive fish such as Loaches and most catfish (Buffering with Triple Sulfa is recommended)
- Do not use with Tetracycline or Erythromycin
- For external use only, do not use products containing Malachite Green for internal use including bath treatment for food fish.

Phenothiazines

Methylene blue (MB) is in the phenothiazine dye class of dyes with numerous applications for human and animal medical use. MB has been used in ruminant animals as an antidote against nitrate and cyanide poisoning. In human medicine, it has been used to treat malaria, depression, and methemoglobinemia and is under current investigation to slow neurodegenerative disease. In aquaculture, MB is effective as an antiseptic and disinfectant, with similar indications for use as MG against *I. multifiliis* and to protect fish eggs from fungal infestation, though with lower efficacy than MG.

Methylene blue

Methylene blue is a heterocyclic aromatic chemical compound with molecular formula: $C_{16}H_{18}ClN_3S$. MB is widely used a Redox indicator in chemistry. Solutions of this substance are blue when in an oxidizing environment, but will turn colorless if exposed to a reducing agent. Since MB is a redox dye and raises the oxygen consumption of cells, this causes the hydrogen oxidized to be passed on to the oxygen. Each molecule of the dye is oxidized and reduced about 100 times per second. Thus, while disinfection results from this, MB is also excellent against methemoglobin intoxication. The therapeutic action of MB on bacteria and other parasites is probably due to its binding effect with cytoplasmic structures within the cell and also its interference with oxidation-reduction processes. Also due to its oxidative reduction properties at therapeutic doses (it is an oxidizing agent occur only at very high doses), MB can also be used as an indicator to determine if a cell alive or not or if the slime coat is healthy or not. The blue indicator turns colorless in the presence of most healthy cells, slime coat, or active enzymes, however, the fish will stain blue where injury has occurred, especially to the slime coat.

Methylene Blue is used for the treatment of following conditions diseases in fish:

- Nitrite poisoning - Fish gasp for breath, tan or brown gills, rapid gill movement known as "gilling"

- Ammonia poisoning - Fish gasp for breath, red or purple gills, lethargic - may lay on the bottom, red streaks on body or fins.

- "White spot" disease or "Ich" caused by *Ichthyophthirius multifilis.*

- Velvet disease caused by a parasite, *Oodinium pilularis.*

- Swim bladder disorder- Fish struggles to maintain proper position, floats upside down, swims with tail over than head

- Fungus infection in eggs- Serves as a safe and mild disinfectant for fish eggs.

- Fish stress - Prophylactic treatment of fish under stress, usually during handling.

Methylene Blue can be used with crustaceans, including crabs, shrimp, and snails, but should be introduced carefully. It will damage live plants and should be used with caution in such cases, being used for only limited periods of time. Methylene blue is useful for:

- A dip/bath for potassium cyanide, ammonia, and nitrite poisoning due to Methylene blue's effect on Methemoglobinemia (nitrite poisoning).

- Antidote for other forms of poisoning including damage to the liver and kidneys caused from poisoning due to being reduced by components of the electron transport chain (a chemical reaction between an electron donor and an electron acceptor to the transfer of H+ ions across a membrane, *via* a set of mediating biochemical Redox reactions).

- Reducing stress during transfer of fish when moving or temporary storage of fish in crowded conditions.

- Treatment of new fish arrivals in a hospital tank, again due to methylene blue's effect on Methemoglobinemia, bacteria, and protozoa.

- As a medicated bath for Dropsy or any other internal malady such as swim bladder problems as MB is easily absorbed by the fish tissue.

- To prevent fungal infection in egg in breeding tanks particularly with bare bottom

- Effective as an antimicrobial swab for sores, tissue damage caused by bacterial pathogens such as Aeromonas or *Flavobacterium* sp., and even as swab following a minor fish surgery to remove tumors or similar.

Contraindications

- Do NOT use Methylene blue with Tetracycline or Erythromycin

- Do NOT use in an aquarium without an established healthy bio filter.

- Do NOT use in the of marine aquarium housed with Anemones, Corals, or Cephlapods (Octopus, etc).

Xanthenes

Xanthene dyes consist of compounds such as fluorescein, rhodamine, and eosin. Compounds from this class are commonly used as fluorescent biological stains and as laser dyes. Rhodamine compounds and fluorescein have been used in tracer studies to monitor the flow of water in rivers and aquatic systems. For example, these dyes were added to pesticide formulations used in sea lice treatment to follow the dispersion of pesticides to surrounding environmental waters. Some dyes from this class have bactericidal, insecticidal, or fungicidal properties. Rhodamine B and the halogenated derivatives such as Bengal and phloxine B showed antifungal action against *Saprolegnia parasitica* in culture studies.

Some dyes from this class act as photosensitizing insecticides. Xanthenes have been formulated for uptake by insects, where they are photoactivated by sunlight to form cytotoxic singlet oxygen and other reactive species. The halogenated eosins (e.g., Rose Bengal, erythrosine, etc.) are effective in this regard. Phloxine B has been commercially developed as a photosensitizing insecticide used to control fruit flies in animal feed. Phloxine B is used to treat the protozoan infection caused by *I. multifiliis* in fish. In this application, phloxine B would be added to an aquaculture pond at night, absorbed by protozoa, and then activated by sunlight to generate free radical species to kill the protozoans. In another study, singlet oxygen produced from the irradiation of Rose Bengal was found to be effective against the virus responsible for white spot syndrome in *kuruma* shrimp populations.

Acridines

Acridine dyes were originally isolated from coal tars and were introduced as an antiseptic in 1912. Acridine dyes such as acriflavine, proflavine, and quinacrine have antiseptic properties. Though not as effective as MG, acriflavine is prescribed for use as a mixture with proflavine to treat external fungal infection in aquarium fish and to disinfect fish eggs.

Acriflavine

An antiseptic agent for the skin and mucous membranes. Generally used for treatment of a fungal infections such as mouth fungus, fin and tail rot saprolegnia, egg fungus (not as strong as Methylene Blue for egg fungus, but safer for main display tank use).

Mildly effective for skin parasites such as Oodinium infection (velvet), sliminess of skin, and ich (although a very mild treatment for Ich, FW or SW). Acriflavin is effective for mild gram-negative bacterial infections. Not generally safe for many crabs, snails, and shrimp at full dose. However Acriflavin can be used with caution in smaller doses with crustaceans and should be buffered and water changes are advised after use.

Azo Dyes

While many azo dyes are regulated in foods as illegal color additives (e.g., Sudan dyes), azo dyes such as Sudan IV (scarlet red) and Congo red are active against Gram-negative bacteria. The azo dye chrysoidine was found to have high bactericidal activity and has been reportedly used to color lower-quality fish to look like more expensive yellowfin tuna.

13

ANTISEPTICS AND DISINFECTANTS

Antiseptics and disinfectants are non-selective, antimicrobial agents which are applied locally to inhibit or kill microorganisms. Their activity ranges from simply reducing the number of microorganisms to within safe limits of public health interpretations (sanitization), to destroying all microorganisms (sterilization) on the applied surface. In general, antiseptics are applied on tissues to suppress or prevent microbial infection. Disinfectants are germicidal compounds usually applied to inanimate surfaces. Sometimes the same compound may act as an antiseptic and a disinfectant, depending on the drug concentration, conditions of exposure, number of organisms, etc. To achieve maximal efficiency, it is essential to use the proper concentration of the drug for the purpose intended.

Antiseptics: These are chemical substances which inhibit the growth or kill micro-organisms on living surfaces such as skin and mucous membrane.

Disinfectants: Disinfection is the process of destruction of pathogenic and other kinds of microorganisms by physical or chemical means. Disinfectants are chemical substances used to destroy viruses and microbes (germs), such as bacteria and fungi, as opposed to an antiseptic which can prevent the growth and reproduction of various microorganisms, but does not destroy them.

Properties of good Antiseptic or disinfectant

- Antiseptics and disinfectants should have a broad spectrum and potent germicidal activity, with rapid onset and long-lasting effect.

- They should not be prone to development of resistance in target microorganisms.

- They should withstand a range of environmental factors (eg, pH, temperature, humidity)

- Must retain activity even in the presence of pus, necrotic tissue, soil, and other organic material.

- High lipid solubility and good dispersibility increase their effectiveness.

- Antiseptic preparations should not be toxic to the host tissues and should not impair healing.

- Disinfectants should be non-destructive to applied surfaces.

- They should be readily biodegradable, not accumulate in the environment, or react with other chemicals to produce toxic residues.

- Offensive odor, color and staining properties should be absent or minimal.

Factors affecting disinfection and antisepsis

Several factors affect the efficacy of antiseptics and disinfectants.

a. **Nature of microbes:** Efficacy of germicide is influenced by types of microorganism. In general, vegetative bacteria are most susceptible to antiseptics and disinfectants, while bacterial spores are more resistant. Among bacteria, gram-positive organism are more susceptible to germicides than gram-negative bacteria.

b. **Size of infection:** Higher the level of microbial contamination, lower is the efficacy of germicides. Presence of large numbers of microorganism at the site of antisepsis or disinfection requires long period of germicidal exposure.

c. **Presence of organic matter:** Presence of blood, pus and tissue debris at the site of action significantly affects the germicidal efficacy of many agents.

d. **Concentration:** Germicidal action of a chemical is inversely dependent on its concentration. For certain compounds, decrease in concentration negatively affects the efficacy, whereas, increase in concentration often produces toxicity.

e. **Temperature and pH:** pH at the site of action may affect compound's efficacy by influencing the compound itself or the microbial cell. Certain compounds are active at alkaline pH. High ambient temperature usually results in increased antimicrobial activity.

Spectrum of activity

The antiseptics disinfectants generally possess wide antimicrobial spectrum, reflecting non-selectivity. In general, the antiseptics are less toxic to microbes than the disinfectants (eg. Alcohols and iodine). It is not generally recommended to use an antiseptic for the purpose of disinfection and *vice-versa*. For disinfection, there is a general scale of innate resistance of microorganisms to germicides.

Mechanism of action

Most of these compounds exert their antimicrobial effect by denaturation of intracellular protein, alteration of cellular membranes (often through extraction of membrane lipids) or enzyme inhibition.

Acids and alkalies

Hydrochloric acid

Hydrogen ion is bacteriostatic at pH ~3–6 and bactericidal at pH <3. Strong mineral acids (HCl, H_2SO_4, etc) in concentrations of 0.1–1 N have been used as disinfectants; however, their corrosive action limits their usefulness. Unionized weak organic acids can readily penetrate and disrupt bacterial cell membranes. Acetic acid, 1% can be used in surgical dressings. At 5%, it is bactericidal to many bacteria.

Sodium Hydroxide

Hydroxyl ion also exerts antimicrobial activity. At a pH above 9, it inhibits most bacteria and many viruses. Hydroxides of sodium and calcium are used as disinfectants. Their irritant or caustic property usually precludes their application on tissues.

A 2% solution of soda lye (contains 94% sodium hydroxide [NaOH]) in hot water is used as a disinfectant against many common pathogens. It is a potent caustic and must be handled with care. Calcium oxide (CaO), ie., lime (hydrated or air-slaked lime), soaked in water produces $Ca(OH)_2$. Aqueous suspensions of slaked lime are used to disinfect premises.

Alcohols

Primary aliphatic alcohols are germicidal. Their potency increases but water solubility decreases with chain length until amyl alcohol (6 carbons) is reached. Antimicrobial effect is related to their lipid solubility (damages bacterial membranes) and their ability to coagulate cytoplasmic proteins. However, they do not destroy bacterial spores. Ethyl alcohol (ethanol) and isopropyl alcohol (isopropanol) are the most widely used alcohols. They can be used in concentrations of 30%–90% in aqueous solutions; best results are usually obtained with 70% ethanol or 50% isopropanol. Higher concentrations tend to be less effective. Isopropanol is slightly more potent than ethanol because of its greater depression of surface tension. "Rubbing alcohol" is a mixture of alcohols, with isopropanol as its principal ingredient. It is used as a skin disinfectant. Alcohol-based hand rinses have rapid-acting antiseptic effects.

Oxidizing agents

Oxidizing agents are chemicals which act by oxidizing the cell membrane of microorganisms resulting in loss of structure, cell lysis and death. Agents those release chlorine and nascent oxygen are strong oxidizers.

Peroxides

These compounds generally exert a short-acting germicidal effect on most organisms through release of nascent oxygen, which irreversibly alters microbial proteins, have little or no action on bacterial spores. Nascent oxygen is rendered inactive when it combines with organic matter.

Hydrogen peroxide solution (3%) liberates oxygen when in contact with catalase present on wound surfaces and mucous membranes. The effervescent action mechanically helps remove pus and cellular debris from wounds and is useful to clean and deodorize infected tissue. The antimicrobial action is of short duration and is limited to the superficial layer of the applied surface because there is no penetration of the tissue. Although its usefulness as an antiseptic is limited, hydrogen peroxide is finding increased application as a disinfectant in water treatment and food processing facilities.

Peracetic acid and the combination of peracetic acid (0.23%) and hydrogen peroxide (7.35%) have been recognized as useful sterilants and antiseptics, combining the broad antimicrobial spectrum and lack of harmful decomposition products of hydrogen peroxide with greater lipid solubility and freedom from inactivation by tissue catalase and peroxidase. They can be used over wide temperature (0°–40°C) and pH (3–7.5) ranges and are not affected by organic matter. They are effective against bacteria, yeasts, fungi, and viruses at concentrations of 0.001%–0.003% and are sporicidal at 0.25%–0.5%. Solutions of 0.2% peracetic acid applied to compresses effectively reduce microbial populations in severely contaminated wounds.

Potassium permanganate has broad antimicrobial properties, but its intense purple color in solution, which stains tissues and clothing brown, is a disadvantage. It is an effective algicide (0.01%) and virucide (1%) for disinfection, but concentrations >1:10,000 tend to irritate tissues. Old solutions turn chocolate brown and lose their activity.

Halogens and Halogen-containing Compounds

Iodine and chlorine are among the oldest topical antimicrobial agents. They owe their activity to high affinity for protoplasm, where they are believed to oxidize proteins and interfere with vital metabolic reactions.

Iodine

Elemental iodine is a potent germicide with a wide spectrum of activity and low toxicity to tissues. A solution containing 50 ppm iodine kills bacteria in 1 min and spores in 15 min. It is poorly soluble in water but readily dissolves in ethanol, which enhances its antibacterial activity.

Iodophores (eg, povidone-iodine and poloxamer-iodine) are combinations of iodine with a solubilizing agent or carrier; they are more stable and water soluble than older formulations. They slowly release iodine as an antimicrobial agent and are widely used as skin disinfectants, particularly before surgery. They do not sting or stain. Iodophores are non-toxic to tissues. They are effective against bacteria, viruses, and fungi but less so against spores. Iodophor solutions retain good antibacterial activity at pH <4, even in the presence of organic matter, and often change color when the activity is lost. Phosphoric acid is often mixed with iodophores to maintain an acidic medium.

The disinfectant action depends on the following reactions:

- The formation of N-iodo radicals on the side chains of basic amino-acids. This prevents hydrogen bonding at these points and so reduces the secondary and tertiary folding of proteins
- The oxidation of –SH groups in cytein, preventing the formation of –S-S- links in proteins
- Reaction with phenolic groups in tyrosine
- The oxidation of unsaturated fatty acids in membranes altering their permeability

All except the last reaction are with poteins and all involve consumption of available iodine. Iodophores are generally inactivated by proteins, so to obtain satisfactory disinfection of surfaces all oganic material and especially proteinaceous material must be removed first. Many iodophores change colour when their bactericidal potency is exhausted.

Chlorine

Chlorine exerts a potent germicidal effect against most bacteria, viruses, protozoa, and fungi through formation of undissociated hypochlorous acid (HOCl) in water at acid to neutral pH. It is effective against most organisms at a concentration of 0.1 ppm, but much higher concentrations are required in the presence of organic matter. Alkaline pH ionizes chlorine and decreases its activity by reducing its penetrability.

Inorganic chlorides include sodium hypochlorite solutions (bleach). A 2%–5% NaOCl solution is a commonly used as an effective disinfectant. Calcium hypochlorite is used as a disinfectant.

Chloramine-T

Chloromnine-T is used as a disinfectant, and as a treatment for bacterial gill disease and occasionally fin-rot. In comparision with formalin, Chloromine-T has greater activity against bacteria but less activity against protozoa. It is regarded as a safer disinfectant than chlorine because it combines with organic matter to form carcinogenic tichloro-methanes. For dip treatments concentrations of 8.5-10 ppm are recommended, with exposure for 1 hour daily for 3 days.

Phenols and related compounds

Phenolic compounds used as antiseptics or disinfectants include pure phenol and substitution products with halogens and alkyl groups. They act to denature and coagulate proteins.

Phenol (carbolic acid) is one of the oldest antiseptic agents. It is bacteriostatic at concentrations of 0.1%–1% and is bactericidal/fungicidal at 1%–2%. The bactericidal activity is enhanced by EDTA and warm temperatures; it is decreased by alkaline medium (through ionization), lipids, soaps and cold temperatures. Concentrations >0.5% exert a local anesthetic effect, whereas a 5% solution is strongly irritating and corrosive to tissues. Phenol has good penetrating power into organic matter and is mainly used for disinfection of equipment or organic materials that are to be destroyed. Because of its irritant and corrosive properties and potential systemic toxicity, it is not used much as an antiseptic currently.

Cresol (cresylic acid) is a mixture of ortho-, meta-, and paracresols and their isomers. It is a colorless liquid; however, after exposure to light and air, it turns pink, then yellowish, and finally dark brown. A 2% solution of either pure or saponated cresol "lysol" in hot water is commonly used as a disinfectant for inanimate objects.

Hexachlorophene (a trichlorinated *bis*-phenol) has a strong bacteriostatic action against many gram-positive organisms (including staphylococci) but only a few gram-negative ones. It is used widely in medicated soaps.

Chloroxylenols are broad-spectrum bactericides with more activity against gram-positive than gram-negative bacteria. They are active in alkaline pH; however, contact with organic matter diminishes their activity.

Reducing Agents

Formaldehyde is a gas, whereas **glutaraldehyde** is oil at room temperature. However, both are readily soluble in water. Their solutions are irritating or caustic to tissues, causing coagulation, necrosis and protein precipitation, but have potent germicidal properties against all organisms, including spores. Their solutions do not lose appreciable antimicrobial properties in the presence of organic matter.

Formalin contains 37% formaldehyde gas in aqueous solution with variable amounts of methyl alcohol to prevent polymerization. A 1%–10% solution of formaldehyde is commonly used as a disinfectant. It is used against many protozoans parasitic on the skin or gills of fish, including *Chilodonella* sp, *Epistylis* sp, *Ichthyobodo necator*, *Ichthyophthirius multifilis* and *Tricodina* sp.

Formalin is a general disinfectant used in hatcheries for the prevention of infection of eggs, the most important being fungi of the genus *Saprolegnia*. The normal use dilution of formalin for fish is 1:6000. Exposure is normally for 30-60 minutes. For eggs, dilution is 1:600, i.e. 10 times the normal concentration for fish.

Glutaral (glutaraldehyde), a 1%–2% alkaline solution (pH 7.5–8.5) in 70% isopropanol, is a more potent germicide than 4% formaldehyde, effective against all microorganisms, including viruses and spores.

14

BLOOD SAMPLING IN FISH

Blood sampling is done in fish for reasons such as haematology, bacteriology, clinical chemistry parameters, parasitological investigations etc. Regular monitoring of fish blood is a very useful diagnostic tool in establishing the health status of fish stocks and to assess the effectiveness of the chemotherapy. Various methods are followed for blood sampling in fish.

a) Severing the caudal peduncle

b) Puncturing the caudal vein

c) Cardiac puncture

d) Dorsal aorta puncture

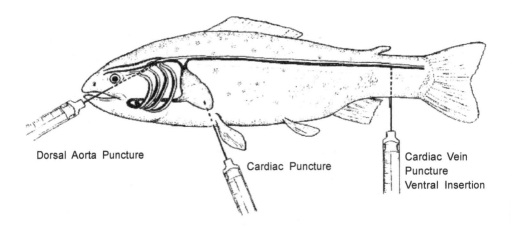

Dorsal Aorta Puncture

Cardiac Puncture

Cardiac Vein Puncture Ventral Insertion

Procedure

It is essential to anaesthetize fish before blood sampling and transfer them to the recovery bath solution.

a. Severing the caudal peduncle

- Ideal sampling method for smaller fish of less than 10 cm (too small to bleed with a syringe or a vacutainer system)

- Anasthetize the fish by immersing the fish in the bath solution with an overdose of anaesthetic so as to sacrifice the fish

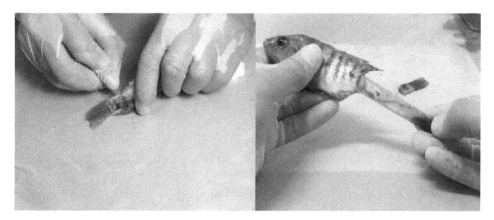

- When the activity of the fish is reduced, severe the caudal peduncle with the scalpel blade or sharp knife. Fill a haematocrit tube with the blood/ or collect the blood with a syringe needle as it flows from the caudal vein.

- The disadvantages of this method are that the fish has to be sacrificed and the blood sample must be collected immediately after the tail is severed as it would coagulate quickly.

b. Puncturing the caudal vein

- Ideal method for repeated blood sampling in larger fish (0.5 to 1 ml blood / 250 g fish / week)

- Insert a needle attached to a syringe or a vacutainer system under the skin of the ventral midline of the caudal peduncle of an anaesthetized or freshly euthanised fish.

- Alternately, insert the needle under the scales of the mid portion of the tail just below the lateral line at a 45° angle to the long axis of the fish.

- Ease the needle toward the vertebral column until you reach the base of the column. Withdraw the needle a fraction of millimetre and obtain the blood sample.

- Remove and discard the needle in a sharps container.

- Return the fish to a recovery bath and monitor the recovery process.

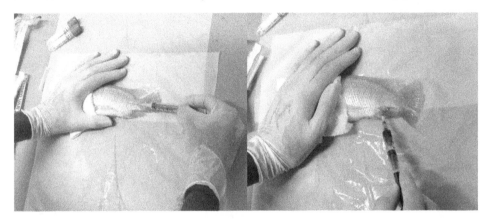

c. **Cardiac puncture:** This is also an ideal method for repeated blood sampling in larger fish. Hold the needle attached to a syringe or a vacutainer system perpendicular to the skin and insert the needle slightly below the tip of the V-shaped notch formed by the gill cover and the isthumus of the fish. Collect the blood as the needle enters the bulbous arteriosis.

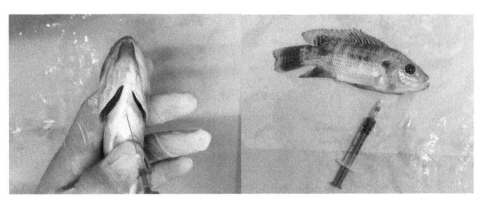

d. Dorsal aorta puncture: This method is suitable for repeated blood collection sample from larger fish. Collect the blood by inserting the needle attached to a syringe or a vacutainer system along the dorsal midline of the mouth just past the juncture of the second gill arch.

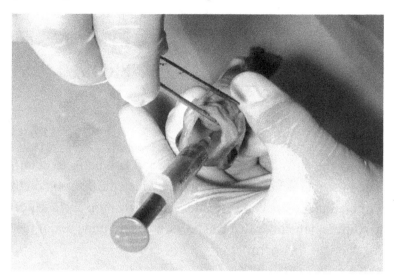

15

SAMPLE COLLECTION FROM FISH

Sampling the eggs and sperm in the brood fish helps to determine the stage of sexual development based on which the decision on the use of hormonal treatment can be made. Catheterization and surgical biopsy are the two different methods that can be used to determine sex of the fish and maturity of the eggs (or sperm).

The ovaries can be sampled with either a rigid or flexible tube (catheter). Rigid catheters are usually made from lengths of glass or hard plastic tubing. Flexible catheters are prepared from lengths of polyethylene or vinyl tubing. The catheters must have an outer diameter small enough to be inserted through the genital opening and sufficient inner diameter to accommodate the eggs. The leading edge of the catheter should also be smoothed or rounded to prevent damage to the fish.

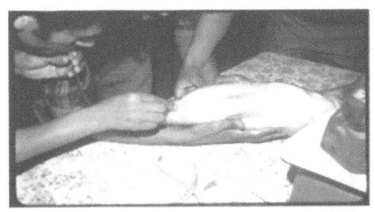

To collect an egg sample, the catheter is inserted through the genital opening and rotated, while gently rotating down the oviduct into the ovary Once the tube is properly inserted, a sample of eggs may be extracted by gentle suction or by removing the catheter while blocking the open end.

www.njfish and wildlife.com

http://www.lssu.edu

The eggs can then be "staged" (their developmental status determined) using a microscope (Rottmann *et al.* 1991). Eggs that are in a late stage of development are uniform in size and large (yolk-filled). They have a germinal vesicle (nucleus) that has migrated away from the center of the egg . A metal or plastic rod with a rounded-conical end and a cavity cut in the rod is used to sample the ovaries of carps. The rounded tip is inserted in the genital opening, and because of its shape, does not puncture the curved oviduct.Before sampling the testes or ovaries, the fish may be anaesthetized, if necessary, with an anaesthetic such as MS-222.Brood fish must be sampled quickly and carefully to minimize physical injury and stress. Egg sample in some fish are also taken by making a small incision along the belly of the fish. First a small amount of physiological solution is drawn into a flexible tube. The tube is inserted through the incision into the ovary and the saline solution is released. Suction is applied to draw a small number of eggs into the tube.The incision is closed with a half-circle surgical needle and suture material and the area is then treated with antibiotic. Mature, conditioned males will often release milt during handling or by applying gentle pressure to the abdomen between the pelvic fins and the vent.

16

ADMINISTRATION OF SUBSTANCES

Substances are administered to fish through a variety of routes. However, the route is determined based on whether the agent is being administered for a local or systemic (Enteral – through digestic tract or parenteral use). The administration method selected should be least stressful to achieve the therapeutic concentration of medication.

Different routes of administration in include

1. Oral - in which the substance may be given into the mouth

2. Intravenous- delivered into the blood vessel

3. Delivery onto, under or across the skin or into the muscle – Epicutneous, Intradermal, sub-cutaneous and intra muscular

4. Intra peritoneal - delivered in the intraperitoneal cavity

5. Gastric gavage - delivered directly into the stomach

6. Branchial diffusion or Inhalation

Administration of substances in fish can be done through

- The rearing medium- ie. Branchial diffusion or Inhalation

- Oral ingestion – generally *via* addition to the diet

- Injections – Intraperioneal, intramuscular

1. Branchial diffusion

Administering through the rearing medium is relatively simple. Hence it is the most common route utilized. Medication through water or bath treatments are very effective against external infections There are three basic types of water medication or bath treatments followed for the treatment of diseases in fish.

a. **Dip treatment:** The fish is dipped into a concentrated chemical bath for a short period of time, usually less than a minute. Prolonged exposure to chemical should be avoided as it would be fatal to expose the sick fish as they will be in stressed condition. Dip treatment in salt water for freshwater fish and in freshwater for marine fish is widely practised for the treatment of ectoparasitic infestation.

b. **Short term bath:** The fish are subjected to a moderate chemical concentration for a period of time ranging from 30 minutes to several hours. This method is widely followed for the treatment of fish. The concentration of the chemical used, the duration of exposure has to be determined. Continuous and vigorous aeration should be provided during the treatment.

c. **Prolonged bath:** The fish are exposed permanently to low concentration of chemical added to the rearing tank. This method is followed for treating the sick fish in ponds.

When administering organic chemicals to fish *via* the water, they have to be dissolved in the carrier solvents such as acetone or methanol. However, the use of carrier solvents in aquatic toxicity testing can be problematic in several ways and use of the solvent requires additional animals/test to assess its direct toxicity or interactions with the test material.

When test material/ chemicals/ substances to be administered are available only in limited quantities, or poorly soluble in water, that necessitates a dietary or injection route of exposure (Kahl et al. 2001).

The exposed fish take up the chemical agents *via* the gills. Fish gills have a large surface area due to a series of lamellae protruding from the surface. The epithelium of the lamellae is extremely thin and designed to facilitate the diffusion of respiratory gases. In addition to gas transfer, fish gills also permit uptake of other molecules. Diffusion or uptake efficiency of chemicals by the gills depends primarily on their hydrophobicity and molecular size

The information on the volume of the water to be treated, the concentration of the active component of the chemical should be known beforehand. The treatment should be terminated and the fish should be transferred to the recovery bath if the fish show any sign of distress. When bath treatments are administered, there should be close observation and maintenance of water quality, as this is a major source of problems. In cases where the anticipated effects are unknown, a small number of fish should be tested before application to the group as a whole.

2. Oral

If a treatment compound is to be administered orally, the volume dose rate should not exceed 1% body weight (1 mL/100 g). Fishes may be force-fed with liquids and semi-solid solutions using flexible rubber tubing and a syringe. Force-feeding is useful for the delivery of stable isotope-labelled compounds and other test substances. In some species, light anaesthesia is necessary to prevent struggling and vomiting. Regurgitation may occur in some species after force-feeding. Fish should be carefully observed particularly following resuscitation, to ensure that the administered agent is retained. However, many fish have a J-or U-shaped stomach or pyloric flexure that prevents the introduced substance from being regurgitated,providing the tube is inserted into the stomach past this flexure.

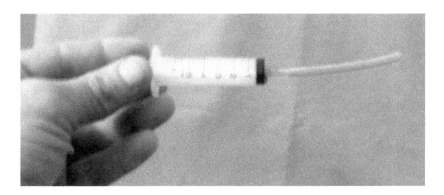

The drugs for the treatment of diseases can also be delivered orally by mixing with the feed at the desired concentration, often coated with the feed binders to avoid leaching. Such medicated feeds are fed to the fishes for a period of time.

3. Injections

Care should be taken during injection to introduce the needle in spaces between the scales. Intramuscular(IM) injections may be made into the large dorsal epaxial and abdominal muscles, taking care to avoid the lateral line and ventral blood vessels. Intraperitoneal (IP)injections should avoid penetrating abdominal viscera as substances that cause inflammation may lead to adhesion formation. The most useful routes for injection in fish are intravascular, intraperitoneal and intramuscular.

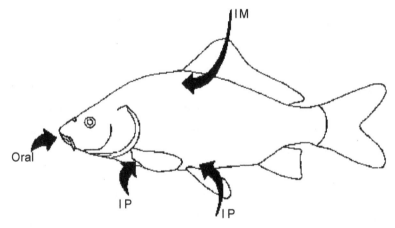

IP – Intraperitoneal, IM- Intramuscular

Chemicals to be injected should be dissolved directly in sterile physiological saline. However, hydrophobic chemicals should be dissolved in very small quantities of co-solvent (e.g., ethanol, methanol, or dimethyl sulfoxide [DMSO]) prior to dilution in saline. For chemicals that are not soluble or stable at neutral pH, the pH of the injection solution may be adjusted with an acid or base. Final injection volumes should be as small as possible to minimize physiological disturbances to the fish. In addition, control fish (vehicle and/or sham injected) should be part of the experimental protocol to correct for any effects of the injection procedure or the vehicle.

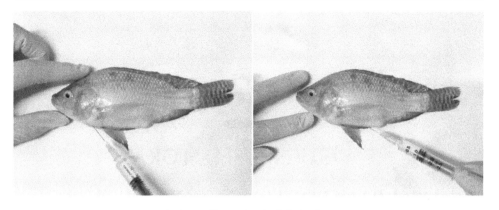

Intraperitoneal administration

Intramuscular administration

Bath treatment

17

DRUGS USED FOR TRANSPORTATION OF FISH

Fishes are easily stressed by handling, transport and stress can result in immuno-suppression, physical injury, or even death. In aquaculture, anesthetics are used during transportation to prevent physical injury and reduce metabolism (DO consumption and excretion). They are also used to immobilize fish so they can be handled more easily during harvesting, sampling and spawning procedures.

An ideal anesthetic should induce anesthesia rapidly with minimum hyperactivity or stress. It should be easy to administer and should maintain the animal in the chosen state. When the animal is removed from the anesthetic, recovery should be rapid. The anesthetic should be effective at low doses and the toxic dose should greatly exceed the effective dose so that there is a wide margin of safety.

Anesthesia is generally defined as a state caused by an applied external agent resulting in a loss of sensation through depression of the nervous system. Anesthetics may be local or general, depending on their application. The method of administration for each anesthesia is fairly well defined, but the appropriate time or circumstance for using it is less clear.

Stages of anesthesia

Induction

Most anesthetics can produce several levels or stages of anesthesia. Stages include sedation, anesthesia, surgical anesthesia and death. The stage achieved usually depends on the dose and the length of exposure. When an anesthetic is first administered (induction) fish may become hyperactive for a few seconds.

Maintenance

Once the desired degree of anesthesia is reached, it may be desirable to maintain fish in that state for some time. Because drug dose and exposure time are often cumulative, it is difficult to maintain a uniform depth of anesthesia. One reason for this is that levels of anesthetic may continue to accumulate in the brain and muscle even after blood levels have attained equilibrium. A desired level of anesthesia can usually be maintained by reducing the dosage. The condition of the animals must be visually monitored during this maintenance period. A change in breathing rate is the most obvious indicator of over-exposure. If this occurs, animals must be moved or the systems flushed immediately.

Recovery

During the recovery stage the anesthetic is withdrawn and fish return to a normal state. To reduce recovery time, induction should be rapid and handling time should be minimal. Initial recovery may take from a few seconds to several minutes, depending on the anesthetic administered. Typically, the animal will begin to respond to noise and other sensory stimuli. Full recovery can take minutes to hours, depending on the species and drug used.

Stages of anaesthesia	Description
I	Loss of equilibrium
II	Loss of gross body movements but with continued opercular movements
III	Cessation of opercular movements

[Table Contd.

Contd. Table]

Stages of anaesthesia	Description
Stages of Recovery	
I	Body immobilized but opercular movements just starting
II	Regular opercular movements and gross body movements beginning
III	Equilibrium regained and pre-anesthetic appearance

(From Iwama *et al.*, 1989)

Before sedation or anaesthesia, fish should be fasted for atleast 12 to 24 hours or until one can ensure that the stomach is empty of food to prevent regurgitation. Anaesthetic agents are inhaled through the gills and rapidly enter the blood. From there they are transported to the central nervous system and excreted *via* the gills. Upon the fish's return to freshwater, a calming effect follows by a successive loss of equilibrium, mobility, consciousness and reflex action.

Rules for anaesthetizing fish

- Use water from which animals originate and at the temperature to which animals are acclimated to anaesthetic solutions.
- Recovery tanks should be aerated at a level required by the particular species of fish.
- Monitor temperature, dissolved oxygen concentration and ammonia levels

An ideal anaesthetic agent should have the following properties

- Produce rapid anaesthesia (1-5 mins) effect
- Have a quick recovery (5 min) time
- Larger safety factor and no/low tissue residue
- Easy to handle and affordable
- No persistent effects on fish physiology and behavior

- Rapid excretion or metabolism and excretion
- Zero or short withdrawal time

Anaesthetic drugs used in aquaculture
MS-222

The chemical name for MS-222 is tricaine methanesulfonate. It comes as a white, crystalline powder that can be dissolved in water. It lowers the pH of water, creating an acidic condition that can irritate fish and cause harmful side effects. To prevent problems, the stock solution can be buffered with sodium bicarbonate (baking soda) to achieve a pH of 7. One of the major drawbacks of MS-222 is that even when fish are deeply anesthetized, handling still increases levels of plasma cortisol concentrations, an indicator of stress. Induction is rapid and can take as little as 15 seconds.

Recovery is usually rapid and equilibrium can be expected to return after only a few minutes. A recovery time longer than 10 minutes suggests that too much anesthetic is being used or that the exposure time is too long. MS-222 has a good safety margin in fish. The drug is more potent in warm waters with low hardness. MS-222 is excreted in fish urine within 24 hours and tissue levels decline to near zero in the same amount of time. It is approved for use on food fish in the U. S. and the United Kingdom, but was recently banned in Canada. The withdrawal time for MS-222 required by FDA is 21 days, which makes it impractical as an anesthetic for fish enroute to market.

Benzocaine

Benzocaine, or ethyl aminobenzoate, is a white crystal that is chemically similar to MS-222. However, benzocaine is almost totally insoluble in water and must first be dissolved in ethanol or acetone. The standard approach is to prepare a stock solution in ethanol or acetone (usually 100 g/L) that will keep for more than a year when sealed in a dark bottle. In solution, benzocaine is neutral (pH 7) and therefore causes less hyperactivity and initial stress reaction than unbuffered MS-222. Benzocaine is effective at approximately the same doses as tricaine (25 to 100 mg/L). It is not safe for exposures longer than 15 minutes. Its efficacy is not affected by water hardness or pH. As with MS-222, it is fat-soluble and recovery times can be prolonged in older fish or gravid females. Benzocaine is not approved by FDA for use on food fish in the U.S.

Quinaldine

Quinaldine is a yellowish, oily liquid with limited water solubility that must be dissolved in acetone or alcohol before it is mixed with water. While it is an effective anesthetic, it is an irritant to fish, has an unpleasant odor, and is a carcinogen. The low cost of quinaldine has made it a popular tool for collecting tropical fish for the aquarium trade, as well as in the bait and sport fish industries. Quinaldine sulfonate is a pale yellow, water-soluble powder; it is more costly than quinaldine or MS-222. Quinaldine solutions are acidic and are usually buffered with sodium bicarbonate. Induction takes 1 to 4 minutes and may cause mild muscle contractions. Recovery is usually rapid. The effective treatment concentration of quinaldine solutions varies with species, but is generally 15 to 60 mg/L. Quinaldine may not produce the deep anesthesia needed for surgery because some reflex responsiveness is usually retained. Fish under full quinaldine anesthesia normally do not stop their gill ventilation so are not as susceptible to asphyxia from respiratory arrest as they are with MS-222. In general, the potency of quinaldine is higher in hard water and warm water. Quinaldine is not approved by the FDA for use on food fish in the U.S.

2-Phenoxyethanol

2-Phenoxyethanol is an opaque, oily liquid. This drug is moderately soluble in water but freely soluble in ethanol. The solution is bactericidal and fungicidal and is, therefore, useful during surgery. It is relatively inexpensive and remains active in the diluted state for at least 3 days. Concentrations of 300 to 400 mg/L are useful for short procedures, and lower concentrations of 100 to 200 mg/L are considered safe for prolonged sedation, such as during transport. 2-Phenoxyethanol is not approved by FDA for use on food fish in the U.S.

Metomidate

Metomidate has been used extensively in human medicine. It anesthetizes fish without the usual stress of an elevated heart rate. Induction is rapid (1 to 2 minutes) and recovery is faster than with MS-222. It anesthetizes salmonids at doses of only 2 to 6 mg/L; low doses are also effective in catfish. In salmonids, metomidate is reported to be more potent in larger, sea-water adapted fish than in freshwater fingerlings. With larval goldfish, *Carassius auratus*, and red drum, *Sciaenops ocellatus*, it has been reported to produce inadequate anesthesia with high mortalities. Metomidate is not approved in the U. S. for use on food fish and is not widely used.

Clove oil

Clove oil has been widely used as an anesthetic in human dentistry and as a food flavoring. The major constituent (70 to 90 percent by weight) is the oil eugenol, but clove oil contains a wide range of other compounds that impart its characteristic odor and flavor. It is an effective anesthesia in carp (*Cyprinus carpio*) at 40 to 120 mg/L. In rainbow trout, *Oncorhynchus mykiss*, doses as low as 2 to 5 mg/L produced sedation sufficient to transport the fish, while doses of 40 to 60 mg/L for 3 to 6 minutes gave effective surgical anesthesia. Recovery time increases with higher doses and longer exposure time. Clove oil is also an effective anesthetic for crustaceans at doses of 100 to 200 mg/L. Clove oil has a very high margin of safety; however, it also requires a relatively long recovery time compared to MS-222. The major advantage of clove oil is that it is inexpensive and not unpleasant to work with. Clove oil is not approved for use on food fish in the U.S.

Aqui-S

Aqui-S is a relatively new anesthetic for fish developed by the Seafood Research Laboratory in New Zealand. This compound is approximately 50 percent isoeugenol and 50 percent polysorbate 80. A dosage of 20 mg/L is effective for most fish species and induction is described as "stress free" because the substance suppresses cortisol. A recent study indicated that Aqui-S was an effective anesthetic on freshwater prawns, but only at much higher concentrations of 100 to 200 mg/L (S. Coyle and J. Tidwell, unpublished data). Currently, Aqui-S is approved for use on food fish in Australia and New Zealand, with no withdrawal period. It is undergoing the New Animal Drug Approval process for use in the U. S., with no withdrawal time. It is used primarily for the "rested harvest" of commercial fish species, where the low stress induction improves the color, texture and appearance of the product. If approved for use in the U. S. with the zero withdrawal time, Aqui-S would be a valuable tool to use when transporting live food fish to market.

Administration of anesthetic agents

Generally, the anesthetic drugs are administered as a single agent in fish. However, in veterinary and clinical practice, combining agents with different properties provides a more complete anesthesia than one single drug alone. Complementary effects between the different agents can result in safer, lower doses. In some cases, induction and recovery are improved and adverse side effects are reduced. Combination anesthesia has been explored in fish. For example, MS-222 and quinaldine administered to rainbow trout (*Oncorhynchus mykiss*) and northern pike (*Esox lucius*) resulted in less mortality and adverse side effects. Improvements in induction and recovery rates were dependent on body size, with smaller fish having quicker induction times but larger fish exhibiting a much faster recovery rate compared with the use of one agent. Thus, combination anesthesia may be safer because this allows a reduction in dose, which is generally reflected in better recovery and lower mortality rates along with reduced adverse side effects in some cases. More experimental studies are needed to develop reliable combination protocols on a greater variety of fish species.

Factors affecting anesthesia

Biological factors such as age, sex, body condition and weight, developmental stage, growth and physiological status, health, and reproductive condition, as well as abiotic factors such as water quality, temperature, and oxygenation affect the efficacy of fish anesthesia. In fish, body condition, water temperature,

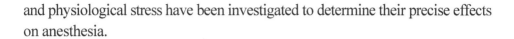

and physiological stress have been investigated to determine their precise effects on anesthesia.

Body Condition

Drug dosing is often relative to the weight of an animal. However, some experimental studies on fish conclude that there is no effect of weight on induction and recovery, whereas others have opposite findings. Larger body size in whitefish (*Coregonus lavaretus*) was found to be associated with decreased induction times; in contrast, larger-sized rainbow trout had longer induction times and there was no effect in Atlantic salmon or brown trout (*Salmo trutta*). Induction increased with greater body weight in Senegalese sole (*Solea senegalensis*) using 2-phenoxyethanol, and metomidate, but not for MS-222.

Water Temperature

Fish are poikilothermic and as such their physiology and metabolic rate are dependent on ambient water temperature. Studies have explored the impact water temperature has on the efficacy of anesthetic agents. Higher temperatures often reduce induction and recovery times. For example, 2-phenoxyethanol in Atlantic cod, Atlantic halibut, European sea bass, and gilthead sea bream has shorter induction and recovery rates at higher temperatures. MS-222 anesthesia is faster at higher temperatures in a variety of freshwater and marine fish. However, there is not a consistent simple relationship with water temperature relative to induction and recovery.

Physiological Stress

Anesthesia is profoundly affected by stress in fish. Stress results in increased cardiovascular responses and gill blood flow, producing greater diffusion of immersion anesthetic agents. Therefore, it is vital that stress is minimized before and during the anesthetic event. Many anesthetic drugs elicit hormonal stress responses as a side effect. Acute stress before anesthesia with MS-222 in Atlantic cod resulted in shorter induction time and prolonged recovery. A deeper

Dose rates of major anaesthetic drugs, evaluated experimentally, for a number of commonly cultured fish species.

Anaesthetic	Atlantic salmon	Rainbow trout	Common carp	Channel catfish	Nile tilapia	Striped bass
MS-222	40-50 mg/L	40-60 mg/L	100-250 mg/L	50-250 mg/L	100-200 mg/L	100-150 mg/L
Benzocaine	40 mg/L	25-50 mg/L	ND	ND	25-100 mg/L	50-100 mg/L
Quinaldine	25-40 mg/L	ND	10-40 mg/L	25-60 mg/L	25-50 mg/L	25-40 mg/L
2-phenoxyethanol	100-200 mg/L	100-200 mg/L	400-600 mg/L	ND	400-600 mg/L	ND
Metomidate	2-10 mg/L	5-6 mg/L	ND	4-8 mg/L	ND	7-10 mg/L
Clove oil	10-50 mg/L	40-120 mg/L	40-100 mg/L	100 mg/L	ND	60 mg/L
Aqui-S	10-50 mg/L	20 mg/L	ND	20-60 mg/L	ND	ND

(ND = Not determined)

plane of anesthesia was observed in these fish after an acute stressor such that the dose of MS-222 was reduced to avoid mortality. However, the benefits of using anesthetic agents during potentially stressful procedures to render the fish unconscious are important to minimize any negative impacts on their welfare. Several studies have shown that handling stress is profoundly reduced when fish are anesthetized. Cortisol is elevated during handling in Atlantic salmon, but when anesthetized with metomidate, cortisol release is prevented. Therefore, the physiological status of the fish should be evaluated before anesthetizing the animal so that one can determine the most appropriate agent and dose. Monitoring of heart rate during prolonged procedures is advisable, because this physiologic parameter is a direct reflection of the fish's level of anesthesia.

Anaesthetics in aquaculture are mainly used for

- Transportation of fish seed and brood fish
- Injection of broodfish
- Treatment of fish for diseases and parasites
- Tagging and marking of fishes

18

IMMOBILIZATION, SEDATION AND ANAESTHESIA IN FISH

Removing fish from water and handling cause stress to them. Rendering fish quiet (sedation) or unconscious (anaesthesia) is crucial for several activities that involve fish handling.Immobilisation / anaesthetization in fish is done for the following reasons

1. Legal requirements

2. Ethical reasons

3. Handling

4. Reduce stress and damage

5. Surgical procedures

6. Blood sampling

7. Marking

8. Killing of fish

9. Endpoint in experiments

10. Animal welfare reasons

Anesthesia substances are mainly absorbed through the gills and also through the skin in fish. Anaesthetic agents inhaled through the gills rapidly enter the blood and transported to the central nervous system and excreted through the gills upon the tansfer of fish to the new rearing medium. They work by inducing a calming effect followed by a successive loss of equilibrium, mobility, consciousness, and reflex action.Respiratory and cardiac failure follows overdose or exposure.

Immobilization of fish is advised for procedures involving handling of fish for a quite short period (few minutes). It could be achieved by submersion in ice water (5 parts ice/1 part water, 0-4° C) for few minutes which reduces the activity of the fish (can be assessed based on the reduced opercular movements).

During transport, sedation (light anaesthesia) of the fish is desirable, since oxygen consumption, CO_2 and NH_3 production are all decreased. However, deep sedation is undesirable because the fish may fall to the bottom, pile up and get injured.It is best to sedate the fish in the holding facility for 30 min before loading and then to continue exposure to a lower concentration of sedative during transport. The use of anaesthetics should not be relied on for increased load carrying capacity. It is not recommended to use anaesthetics on small fish transported on small distances, since in such conditions the space factor has a greater influence on the health of the fish than the accumulation of metabolic products. MS-222 is a very mild tranquilizer and fish easily recover from its effects at recommended to rate of 20 mg/l for carp and grass carp, 10 mg/1 for silver carp, and 35 mg/1 for bighead carp. At these concentrations the fish can still hold their natural position but their respiration and motility are significantly decreased.

However, for procedures involving handling of fish for a long period anasthetization of fish is essential. Addition of anaethetics to water is similar to inhalation anaesthesia given to terrestrial animals. The drug is taken up through water and enters the arterial blood and the excess is secreted through gills, kidney and skin.

Guidelines for anaesthesia of fish (Ross and Ross, 1999)

- The fish should be starved 12 to 24 hours prior to anaesthesia to avoid regurgitation of food and associated water quality problems
- The anaesthetic bath and the recovery bath should contain water originating from the same source to avoid variations in the water quality parameters (Check water quality -pH, O_2, temperature)

- Aerate the bath using an air diffuser or air stone
- Fish should be anesthetized in small batches and remain no longer than ten minutes in the anaesthetic solution

Depth of anaesthesia can be assessed by

a. Ataxia, loss of reflex and response to stimuli (squeezing the base of the tail)

b. Observing the movement of operculum (gill cover) / respiratory rate

c. Monitoring the heart rate using a Doppler or /and ECG (using pectoral and anal fins)

d. Observing the color of gills – should be light red or pale , suggestive of hypoxemia, hypotension or anaemia

Most anesthetics can produce several levels or stages of anesthesia as shown below. The stage achieved usually depends on the dose and the length of exposure. When an anesthetic is first administered (induction), fish may become hyperactive for a few seconds and undergo the following stages.

Stage	Condition	Behavior/Response
I	Sedation	Motion & breathing reduced
II	Anesthesia	Partial loss of equilibrium Reactive to touch stimuli
III	Surgical anesthesia	Total loss of equilibrium No reaction to touch stimuli
IV	Death	Breathing & heart beat stops Overdose - eventual death

A wide range of water soluble anaesthetics are used in fish to minimize stress to fish while out of water; to reduce pain and to immobilize fish during a procedure. The essential characteristics of an ideal anaesthetic are

1. It should induce anaesthesia rapidly with minimum hyperactivity or stress.

2. It should be easy to administer and should maintain the animal in the chosen state.

3. Recovery should be rapid when the animal is removed from the anaesthetic.

4. It should be effective at low doses and the toxic dose should greatly exceed the effective dose so that there is a wide margin of safety

    - Fish are usually anesthetized by immersing them in an anesthetic bath containing a suitable concentration of drug so that the drug is absorbed through the gills and rapidly enters the blood stream.

    - Applying an anesthetic solution to the gills with a spray bottle can be useful with large animals or if immersion is impractical.

    - Continous irrigation of the anesthetic solution over the gills is followed if the fishes have to be kept out of water for a prolonged period as during surgery.

19

DRUGS USED IN THE TREATMENT OF BACTERIAL AND VIRAL DISEASES IN FISHES

In every sphere of production and productivity, sustainability is an essential requirement for long term viability. Therefore, the hallmark of success of any production processess whether it is agricultural or industrial is determined by its capability to maintain its sustainability over the years. In aquaculture production, the concept of sustainability has attempted from time immemorial, but has not been achieved, as the production is dependent on various extrinsic and intrinsic factors. Being a biological process that is carried out with full dependence on natural environment and the aquatic animals cultured, disease is a major problem in production systems. In order to overcome the the stress and disease situations practioners are forced to use a variety of drugs, probiotics and chemicals.

Drugs

Aquaculture drugs are chemicals or biological preparations or formulations which can be used in aquaculture systems to be administered to aquatic organisms at any stage of their life for management of their health.

Mode of action

- Produce a psychological response upon the host fish
- Destroy or overcome the agents responsible for infection

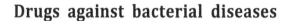

Drugs against bacterial diseases

1. Florfenicol

2. Hydrogen peroxide

3. Sulphadimethoxine

4. Oxytetracycline hydrochloride

5. Oxytertacycline dehydrate

I. Florfenicol

Species	Indication	Dosing	Limitations & comments
Catfish	Control of mortality due to enteric septicaemia associated with *Edwardsiella ictaluri*	10 mg /kg fish per day for 10 days	12 day withdrawal time
	Control of mortality due to columnaris disease associated with F*lavobacterium columnare* (conditional approval) Must use		
Fresh water reared salmonid	Control of mortality due to furunculosis associated with *Aeromonas salmonicida*	10 mg /kg fish per day for 10 days	
	Control of mortality due to cold water disease associated with *F. psychrophillum*		

2. Hydrogen peroxide

Species	Indication	Dosing	Limitations & comments
Fresh water reared finfish eggs	Control of mortality due to Saprolegniasis	500-1,000 mg/L for 15 min in a continuous flow system once daily on consecutive alternate days until hatching	Initial bioassay on small number of fish is recommended before treating the entire group

[Table Contd.

Contd. Table]

Species	Indication	Dosing	Limitations & comments
		Warm water: 750-1,000 mg/ L for 15 min in a continuous flow system once daily on consecutive alternate days until hatching	0 day withdrawal time
Fresh water reared salmonid	Control of mortality due to bacterial gill disease associated with *Flavobacterium branchiophillum*	100 mg/L for 30 min or 50 mg/L for 60 min once per day on alternate days for three treatments	Initial bioassay on small number of fish is recommended before treating the entire group 0 day withdrawal time
Freshwater reared coldwater finfish and channel catfish	Control of mortality due to external columnaris disease associated with *F. columnare*	Fingerlings and adults: 50-75 mg/L for 60 min once per day on alternate days for three treatments Fry: 50 mg/L for 60 min once per day on alternate days for three treatments	Initial bioassay on small number of fish is recommended before treating the entire group

3. Sulfadimethoxine & Ormethoprim

Species	Indication	Dosing	Limitations & comments
Salmonids	Control of furunculosis due to *Aeromonas salmonicida*	50 mg/Kg fish per day for 5 days	In feed, 42 days withdrawal time
Catfish	Control of enteric septicemia due to *Edwardsiella ictaluri*	50 mg/Kg fish per day for 5 days	In feed 3 day withdrawal time

4. Oxytertacylin hydrochloride

Species	Indication	Dosing	Limitations & comments
Finfish fry and fingerlings	Mark skeletal tissues	200-700 mg/L of water for 2-6hrs	None

5. Oxytertacylin dihydrate

Species	Indication	Dosing	Limitations & comments
Pacific salmon	Mark skeletal tissues	250 mg/Kg fish per day for 4 days	Salmon <30g size in feed as sole ration 7 day withdrawal time
Salmonids	Control of ulcer disease (*Hemophilus piscium*) furunculosis (*Aeromonas salmonicida*) bacteria haemorrhagic septicemia (*A. liquifaciens*) and Pseudomonas disease (*Pseudomonas* spp.)	2.5-3.75 g per 100 lbs fish per day for 10 days	In mixed ration 21 day withdrawal time No temperature restrictions on use
Freshwater reared salmonids	Control of mortality due to cold water diseas caused by *Flavobacterium psychrophillum*	3.75 g per 100 lbs fish per day for 10 days	In mixed ration 21 day withdrawal time No temperature restrictions on use
Freshwater reared *Oncorhynchus mykiss*	Control of disease due to columnaris diseae (*F. columnare*)	3.75 g per 100 lbs fish per day for 10 days	In mixed ration 21 day withdrawal time No temperature restrictions on use
Catfish	Control of bacteria haemorrhagic septicemia (*A. liquifaciens*) and Pseudomonas disease (*Pseudomonas* spp.)	2.5-3.75 g per 100 lbs fish per day for 10 days	In mixed ration 21 day withdrawal time Water temperature not below 62° F (16.7° C)

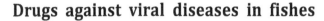

Drugs against viral diseases in fishes

Viruses may have two modes of transmission ie. vertical and horizontal as latent and virulent forms. Therefore, antiviral preparations can be designed in two ways, one which does not permit latent virus to express itself and manifest disease and the second which either blocks viral binding sites to the target cell receptor or disrupts viral protein or nucleid acids. Several compounds which include carbohydrates such as sulphated polysaccharides, lipids such as sulpholipids, proteins such as Cyanovirin N etc. sourced from different families of algae, sponges and bacteria reportedly possess antiviral activities without having much information on the mode of action. Their activity against viruses pathogenic to human beings has been registered but it yet to be scientifically documented against aquaculture pathogens.

Drugs are not effective; however, antibiotics and other drugs may be used to control secondary bacterial infections. Management techniques that minimize stress and crowding, biosecurity measures, temperature manipulation hold the greatest promise for control of piscine viral diseases. An antiviral drug in aquaculture still remains as an unexplored area. Drugs have to be developed to prevent a latent virus from entering into virulence, block virus absorption to the host cell and uncoating, block any stage of virus multiplication and assembly and enhance non-specific defense system so that an invading virus could be easily destroyed.

20

Common Chemicals and their use in Fish Farms

Chemicals are used in aquaculture for various purposes (FAO, 1987)

Most chemicals used for controlling disease organisms are toxic and/or irritant for the skin and respiratory tract. Many chemicals can cause serious health problems if swallowed or absorbed through the skin. You should therefore handle chemicals with care, mark them clearly and store them safely, keep away from children in particular.

When handling chemicals, the following precautions need to be taken :

- protect your hands by wearing rubber gloves; if possible wear overalls and boots for general protection;
- work in a well-ventilated area;
- handle chemicals carefully, avoiding splashing or spillage and, if accidentally splashed, wash off immediately;
- avoid inhaling dangerous vapours;
- clearly mark all containers, equipment and protective clothing used for storing and handling chemicals; unless they can be thoroughly and safely washed out, do not use them for other purposes; store them safely when not in use;

- thoroughly wash your hands after use, and particularly before touching any food.

Most chemicals degrade with time. When buying chemicals ensure that they are still of good quality. Check their expiration or "sell by" date, if any. Check also whether they have been stored properly. Clearly indicate on each container the date of purchase.

For storing chemicals, use a cool, dark, dry and lockable room. Some chemicals may require refrigerator storage; check labels or handling instructions. Generally, chemicals deteriorate more quickly in warm conditions. Keep good records and ensure a good rotation of the stocks.

Determining the strength of chemicals and their solutions

Treatment chemicals are costly and can, in the wrong dosage, be toxic to fish. It is therefore essential to know and understand how the strength of these chemicals is expressed and how dosages can be calculated. You should become familiar with concentrations of chemical solutions, treatment units and their conversion values. You will then be able to avoid wasting chemicals and losing fish.

Treatment chemicals contain one or more toxic ingredients to disable or kill disease organisms. These are called the active ingredients (Al). The amount of active ingredient contained depends on the chemical. The Al concentration is expressed in percent of the total weight or volume of the chemical.

Examples

- common salt contains 100 percent active ingredient;
- chlorine bleach powder contains 33 percent Al (chlorine)
- Wescodyne contains 1.6 percent Al (iodine);
- Roccal contains from 10 to 50 percent Al (benzalkonium chlorides).

The usual practice is to dilute solid or liquid chemicals in water and prepare :

- Stock solution, which can be kept for some time, and be diluted as required to a working solution or a treatment dosage (also used when you need to dose very small amounts of chemicals);
- Working solution which is used directly once prepared;

- Treatment dosage which is the concentration to which fish are exposed for treatment.

These solutions are prepared to a recommended strength or concentration, expressed either as

- the concentration of the chemical required (e.g. diluted in water, or mixed with feed); or

- the concentration of active ingredient required.

Be very careful to make sure whether it is the chemical or the active ingredient which is being referred to. Examples are given later for making up these solutions. The concentration of material (chemical or active ingredient) in a solution may be expressed in various ways :

(a) As the weight present in the solution volume, such as milligrams per litre solution (mg/l), grams per cubic metre (1000 l) solution (g/m³), grams per 25 l solution (g/25 l);

(b) As a volume present in the solution volume such as milligrams per litre solution (ml/l), millitres per cubic metre solution (ml/m³), millitres per 10 l solution (ml/10 l).

In practice, you may consider that :

| 1ml water weighs 1 g |
| 1 l water weighs 1 kg |

(c) As a percentage, the number of parts (normally be weight) of material in 100 parts of solution.

Example

- A 1.5 percent solution of salt contains 1.5 g salt per 100 g = 100 ml solution or 15g/l; a 5 percent solution of chlorine bleach powder contains 5 g chlorine per 100g = 100 ml solution or 50 g/l.

  In parts per million (ppm), the parts of material present in one million parts of solutions; by weight this is equivalent to mg/l, g/m³, or by volume to ml/m³ of solution.

- A 1000 ppm solution of chlorine bleach powder is in fact a 1000 mg/l solution; this amounts to 1 g chlorine per litre; a 160 ppm solution of formalin contains 160 mg/m³ of formalin; this is equivalent to 0.16 ml formalin per litre solution.

(d) As a ratio, the parts of solution per each part of Al, such as a 1:4000 solution where for example each ml Al is present in 4000 ml solution (1 ml/4l)

To obtain a ppm value from a ratio, multiply the ratio by 1000000 as shown in the chart below.

Ratio	ppm	Ratio	ppm
1:500	2000	1:10000	100
1:1000	1000	1:15000	67
1:4000	250	1:100000	10
1:5000	200	1:200000	5
1:6000	167	1:500000	2

Calculating the amount of chemical to be used

If you are using a chemical with 100 percent Al, When the concentration is given as a weight or volume of Al per volume of solution, multiply this concentration by the total volume of water to be treated.

Examples

- A concentration of 3 mg/l malachite green is required as a treatment dosage. Your tank contains 2 m^3 = 2000 l. You will need 3 mg/l x 2000 l = 6000 mg = 6 g malachite green.

- A stock solution of 1 g/l copper sulphate is to be made, to use in a 5-l drum. You will need 1 g/l x 5 l = 5 g of copper sulphate.

  - When the concentration is given as a percentage, multiply this concentration (expressed as a decimal number) by 1000 times the water volume (in l) to obtain the amount of chemical in ml or g.

- Recommended treatment dosage is 2 percent common salt. Your plastic barrel contains 30 l water; you will need 0.02 x 1000 x 30 l = 600 g salt.

  - When the concentration is given in parts per million (ppm), multiply this concentration by the water volume (in l). Divide the result by 1000 to obtain the amount of chemical in ml or g.

- Recommended dosage is 100 ppm copper sulphate. Your trough contains 500 l water. You will need (100 ppm x 500 l) x 1000 = 50 g copper sulphate.
  - When the concentration is given as a ratio, multiply this concentration by 1000 times the water volume (in l) to obtain the amount of chemical in ml or g.
- Recommended concentration is 1:6000 formalin; your tank contains 2000 l water; you will need (1x 6000) x 1000 x 2000 l = about 44 ml of formalin.
- If you are using a chemical that contains less than 100 percent active ingredient, divide the result obtained from the above calculations by the percentage of Al (expressed as a decimal) in the chemical.

Example

Recommended dosage of an insecticide containing 80 percent Al is 0.5 ppm. Your pond contains 50 m^3 = 50000 l water:

- calculate amount of Al required as (0.5 ppm x 50000 l) x 1000 = 25 g;
- calculate amount of chemical required to provide this as 25 g x 0.80 = 31.25 g.

Measuring chemicals

- To weigh chemical powder, crystals, etc., you require a balance with good accuracy within a particular range.
- For small quantities, you may use, for example, a simple microbeam balance or a small spring balance, with good accuracy within the range of 1 to 125 g.
- For larger quantities, you will need a good beam or spring balance similar to the one used to weigh fish accurate within the range of 100 g to a few kilograms, depending on the volumes of water you plan to treat.

- For quantities smaller than 1 g, use stock or working solutions.

    – To measure volumes of liquid chemicals, you require a series of graduated plastic cylinders, with a typical range of capacity from 10 ml to 500 ml. The smaller the individual capacity of the cylinder, the more accurate the measurement.

    – For small volume measurements of one or a few milliliters, it is best to use either a medical syringe (capacity 2 to 5 ml) or a graduated or bulb-type pipette. Obtain these from a local pharmacy, drugstore or medical office.

Using stock and working solutions

- The commonly available balances and measuring cylinders are not accurate enough to measure very small amounts of chemicals such as fractions of a gram or milliliter. It is best then to prepare a stronger stock solution, from which the final solution is prepared. The stock solution can usually be stored, to use as required. In some cases, you can dilute the stock solution to a working solution, which can be used for a short time, for example during one or two days, to be diluted as needed.

- Stock solutions and working solutions are usually made up to concentrations which make them convenient to use, for example :

    – for a stock solution of 1 g (1000 mg) of chemicals per litre, 1 ml contains 1 mg. Thus 1 ml mixed in 1 litre gives a solution of 1 mg/l, whereas 20 ml mixed in 1 litre gives a solution of 20 mg/l, etc.;

    – for a working solution of 10 mg/l (for example made by mixing 10 ml of the previous stock solution in 1 litre of water), 100 ml contains 1 mg and 50 ml contains 0.5 mg of chemical. Thus 50 ml of working solution mixed in 1 litre of water gives a chemical concentration of 0.5 mg/l.

Preparing stock and working solutions

- You can either :

    – use recommended concentrations as suggested by suppliers; or

- make standard strengths as described above; or
- adapt the solutions according to your working needs.

Examples

- You are treating fish stock in 50-l buckets, and you need a treatment dosage of 1 mg/l. You have a 20-ml beaker for dosing the chemical. What strength of solution do you need to give the correct dosage, using one beaker full for every bucket?

- The 50 l bucket requires 50 l x 1 mg/l = 50 mg of treatment chemical. Thus you need 50 mg in each 20 ml beaker. Concentration of required solution is therefore 50 mg / 20 ml = 50 mg x (1000 ml x 20 ml) = 2500 mg/l or 2.5 g/l.

  Note : Do not try to make concentrations stronger than the solubility limits of the chemical; the supplier will have this information. With dissolving solids, you can visibly check the solubility limit as excess solids will remain suspended in the liquid. In general, however, you should try to make the stock solution quite concentrated, as it will usually keep better and it will be less affected by accidental contamination.

Storing and dispensing stock and working solutions

- Stock and working solutions should be stored in clean bottles, preferably made of dark glass. You may also darken a clearer glass or plastic bottle with black tape. Label the bottle clearly, indicating the kind of chemical, its date of preparation, its strength and notes for its use, for example : malachite green stock solution, prepared 6/6/90, 100 g/l – use 2 ml/100 l to make 2 mg/l solution.

- Some chemicals react with other materials, particularly cheap household plastics. Do not use these materials if possible.

- Bottles should be well stopped preferably with a glass, rubber or cork top that will not react with the ingredients.

- Store the chemicals in a cool place, refrigerated if necessary.

- When measuring out the stock or working solution, pour a sample in a clean glass beaker to make sure it has not deteriorated. If acceptable, pour out approximately the amount required, then measure it exactly, using a clean pipette, syringe or measuring cylinder.

- Alternatively, draw out the required amount with a clean pipette or syringe, checking to ensure its quality is still acceptable.

- As unused working solution might have become contaminated by the measuring equipment or by exposure to the open air, it is preferable not to return it to the stock solution.

Some common chemicals for fish farmers

- There are several chemicals which are commonly used by fish farmers to prevent and cure fish diseases.

- **Limes and calcium cyanamide** are particularly useful to control pests in drained ponds.

- **Agro-industrial by-products** such as rice bran, molasses and tobacco dust shavings can be used for similar purposes.

- Organic **poisons** such as rotenone and saponin can control pests in undrained ponds.

- **Household bleach** consists of a weak solution of sodium hypochlorite, which can be used as a good, general disinfectant of non-metallic equipment and working areas. As the active chlorine readily evaporates, a fresh solution should be used within two days.

- **Chlorine bleach liquid** is a stronger solution containing 13 percent active chlorine. It can be diluted or used directly as a strong disinfectant, for example for sterilizing tanks.

- **Chlorine bleach powder** contains 33 percent active chlorine. It makes a very strong disinfectant, and is especially useful for tanks.

- **Iodophores** are organic iodine compounds sold as a brownish liquid under various trade names such as **Wescodyne, Romeiod, FAM 30 and**

Common chemicals to prevent and cure fish diseases

Chemical	Solid or liquid	Earth ponds	Tanks	Equipment	Live stock	
					Eggs	Juveniles / adults
Quicklime	S	P	–	–	–	–
Hydrated lime	S	P	–	–	–	–
Calcium cyanamide	S	P	–	–	–	–
Agro-Industrial by-products	S	P	–	–	–	–
Organic poisons	S	P	–	–	–	–
Household bleach	S	–	P	P	–	–
Chlorine bleach liquid	S		P	P	–	–
Chlorine bleach powder	S	–	P	P	–	–
Iodophores	L	–	P	P	P	–
WescodyneR, RomeiodR, FAM 30R, etc.						
Benzalkonium chlorides	L/S	–	P	P	–	–
RoccalR, HyamineR, etc.						
Common salt	S	–	–	–	–	–
Formalin	L	–	–	–	–	C
Malachite green	S/L	–	–	–	P/C	–
Potassium permanganate	S	–	P	–	–	–
Copper sulphateInsecticides	S	–	–	–	–	P/C
BromesR, DipterexR, NeguvonR, etc.	S	–	–	–	–	C

(**Source** : FAO, 1997   S - Solid, L – Liquid, P – Prevention, C- Cure)

Buffodine. They are excellent disinfectants, but highly toxic to fish. The active ingredient content varies from one make to another, so check carefully before use. A concentrated solution can be stored for several months in a cool, dark place. The diluted solution remains active for up to a week, until it fades from a brown to straw colour.

- **Benzalkonlum chlorides** are a blend of quaternary ammonium compounds generally sold either as a powder or as a solution under various trade names such as **Roccal** (10 to 50 percent Al) and **Hyamine 3500** (50 percent Al). These are very good disinfectants which can be reused for up to one week.

- **Common salt** as used in the kitchen is usually a cheap and easily available chemical (sodium chloride). In solution, it not only kills several disease causing organisms but may also have positive effects on the fish by stimulating appetite and increasing mucus secretion, improving resistance to handling. Depending on the species, excess levels may however stress the fish. Thus cyprinids are more susceptible than salmonids.

- **Formalin** is the commercial name for a 35 to 40 percent solution of formaldehyde gas in water. Formalin is toxic to fish particularly in soft water. As it lowers dissolved oxygen levels, make sure treatment water is well oxygenated. Toxic and irritant for the eyes and lungs, it should be handled very carefully in a well-aerated place. It is very sensitive to light so should be stored in a dark bottle. Always check to see if there is a white deposit at the bottom of the bottle. In such cases, before use, carefully filter out this highly toxic deposit of paraformaldehyde. Keep formalin away from metallic equipment. Remember when determining how much to use that it is considered to be a **100 percent Al chemical.**

- **Malachite green** is sold either as a blue-green to green crystalline powder or as a solution of various strengths. You should use **the medical-grade, zinc-free quality** to avoid heavy fish losses. Do not use in the presence of zinc or iron. Handle carefully and avoid any contact with your skin. If possible, use a trial batch, as quality and toxicity may vary greatly from one batch to another. Not to be used on food fish.

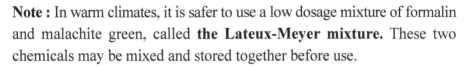

Note : In warm climates, it is safer to use a low dosage mixture of formalin and malachite green, called **the Lateux-Meyer mixture.** These two chemicals may be mixed and stored together before use.

- **Potassium permanganate** is a violet, crystallized powder. It is a good disinfectant in the absence of organic materials that destroy it. When in solution, it should be kept in a dark bottle.

- **Copper sulphate** is sold as a light-blue powder which readily dissolves in water; blue crystals are also common, but they should be small enough to be easily soluble; it is relatively cheap, but is highly toxic for humans and fish. It should be stored safely and handled properly.

- **Agricultural insecticides** are usually organophosphates and are sold under various trade names such as **Bromex, Dipterex, Dylox, Flibol, Masoten and Neguvon.** Carefully check the percentage of active ingredients present. They are often very toxic for various other aquatic organisms including zooplankton and for humans and domestic animals. In ponds, they normally decompose within a few days. Toxicity to disease organisms is often reduced when water temperature exceeds 30° C.

The toxicity of chemicals used

- Always remember that the toxicity of these chemicals for fish may vary greatly according to water quality. In general toxicity increases :

  - as water temperature increases, especially above 25° C;

  - as pH decreases and water becomes acid;

  - as total alkalinity decreases, especially below 50 mg/l $CaCO_3$;

  - as dissolved oxygen concentration reaches below 5 mg/l.

- Check your water quality and look at the chart that follows for the chemical you plan to use. You will learn more about toxicity when considering the choice of disease treatment

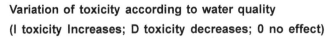

Variation of toxicity according to water quality
(I toxicity Increases; D toxicity decreases; 0 no effect)

Chemical	Water	pH	Tot. alk.	Organic matter
Formalin	I	D	D	-
Copper sulphate	O	D	D	D
Malachite green	O	-	-	-
Pottasium permanganate	I	D	-	D
Insecticides	D	-	-	-

Note: As the water quality factor increases, the toxicity varies as indicated.

REFERENCES

1. Armstrong, S.M., Hargrave, B.T. and Haya, K. (2005) Antibiotic use in finfish aquaculture: modes of action, environmental fate and microbial resistance. Handbook of Environmental Chemistry, 5: 341–357.

2. Boyd, E. and McNevin, A.A. (2015) Aquaculture, Resource use, and the Environment. John Wiley & Sons, Inc. pp. 366.

3. Burridge, L.E., Weis, J.S., Cabello, F., Pizarro, J., and Bostick, K. (2010). Chemical use in salmon aquaculture: a review of current practices and possible environmental effects. *Aquaculture* 306, 7–23.

4. Chopra, I. and Roberts, M. (2001) Tetracycline Antibiotics: Mode of Action, Applications, Molecular Biology, and Epidemiology of Bacterial Resistance. Microbiology and Molecular Biology Reviews, 65 (2): 232-260.

5. Coyle, S.D., Durborow, R.M. and Tidwell, J.H. (2004) Anaesthetics in aquaculture. SRAC publication no. 3900.

6. Fair, R.J. and Tor, Y. (2014) Antibiotics and bacterial resistance in the 21st century. Perspectives in medicinal chemistry, 6: 25–64.

7. FAO/OIE/WHO. (2006) Report of a Joint FAO/OIE/WHO Expert Consultation on Antimicrobial Use in Aquaculture and Antimicrobial Resistance. Geneva, June 2006.

8. Lunestad, B.T., and Samuelsen, O. (2008) Veterinary drug use in aquaculture. In: Improving Farmed Fish Quality and Safety, (eds) O. Lie. Woodhead Publishing, Cambrirge, MA. pp.97–127.

9. Noga E.J. (2010) Fish Disease: Diagnosis and Treatment. Wiley Blackwell. pp.536.

10. Pandey, G. (2017) Fish Pharmacology and Toxicology: Research Reviews. Daya Publishing House. India. pp.220

11. Romero, J., Feijoo, C.G. and Navarrete, P. (2012) Antibiotics in Aquaculture – Use, Abuse and Alternatives, Health and Environment in Aquaculture. Health and Environment in Aquaculture, 159-198

12. Sandhu, H.S. and Rampal, S. (2013) Essentials of Veterinary Pharmacology and Therapeutics. 2nd Ed. Kalyani Publishers. pp.1529.

13. Serrano, P.H. (2005) Responsible use of antibiotics in aquaculture: FAO Technical Paper 469. FAO. Rome, Italy.

14. Svobodova, Z. and Vykusova, B. (1991) Research Institute of Fish Culture and Hydrobiology (Vodnany, Czechoslovakia) and FAO International Training Course on Fresh-Water Fish Diseases and Intoxications: Diagnostics, Prophylaxis and Therapy. Diagnostics, prevention and therapy of fish diseases and intoxications: manual for International Training Course on Fresh-Water Fish Diseases and Intoxications: Diagnostics, Prophylaxis and Therapy. Research Institute of Fish Culture and Hydrobiology, Vodnany, Czechoslovakia.

15. Treves-Brown, K.M. (2000) Applied Fish Pharmacology. Aquaculture, vol 3. Springer, Dordrecht. pp.309.

16. U.S. Food and Drug Administration. (2018) Approved Aquaculture Drugs. Available from:https://www.fda.gov/animalveterinary/development approvalprocess/aquaculture/ucm132954.htm.

17. U.S.Food and Drug Administration. (2018) The drug development process. Available from: https://www.fda.gov/forpatients/approvals/drugs/